WHAT CAN I DO NOW?

Preparing for a Career in Engineering

Ferguson Publishing Company, Chicago, Illinois

Printed in the United States of America

V-4

Library of Congress Cataloging-in-Publication Data

Preparing for a career in engineering
     p. cm. -- (What can I do now?)
  Includes bibliographical references and index.
  Summary: Explores career opportunities in the field of engineering, provides a detailed look at eight specific occupations, discussing education and training needed, skills required, and salary ranges, and offers advice on steps to prepare for a career.
  ISBN 0-89434-248-7
  1. Engineering--Vocational guidance--United States--Juvenile literature. [1. Engineering--Vocational guidance.
  2. Vocational guidance.] I. Series.
  TA157.P72   1998
  620'.0023--dc21   98-7383
                                      CIP
                                      AC

Ferguson Publishing Company
200 West Madison, Suite 300
Chicago, Illinois 60606
800-306-9941
www.fergpubco.com

About the Staff

- Holli Cosgrove, *Editorial Director*
- Andrew Morkes, *Editor*
- Veronica Melnyk, *Assistant Editor*
- Mathew Hohmann, Veronica Melnyk, Beth Oakes, Kate Quinlan, Elizabeth Taggart, *Writers*
- Connie Rockman, MLS; Alan Wieder, *Bibliographers*
- Patricia Murray, Bonnie Needham, *Proofreaders*
- Joe Grossmann, *Interior Design*
- Parameter Design, *Cover Design*

Contents

Introduction

If you're considering a career in engineering—which presumably you are since you're reading this book—you must realize that the better informed you are from the start, the better your chances of having a successful, satisfying career.

There is absolutely no reason to wait until you get out of high school to "get serious" about a career. That doesn't mean you have to make a firm, undying commitment right now. Gasp! Indeed, one of the biggest fears most people face at some point (sometimes more than once) is choosing the right career. Frankly, many people don't "choose" at all. They take a job because they need one, and all of a sudden ten years have gone by and they wonder why they're stuck doing something they hate. Don't be one of those people! You have the opportunity right now—while you're still in high school and still relatively unencumbered with major adult responsibilities—to explore, to experience, to try out a work path. Or several paths if you're one of those overachieving types. Wouldn't you really rather find out sooner than later that you're not as interested in chemical engineering as you thought? That maybe you'd prefer to be a plastics engineer, or a transportation engineer, or a biomedical engineer?

There are many ways to explore the engineering industry. What we've tried to do in this book is give you an idea of some of your options. The chapter "What Do I Need to Know About Engineering" will give you an overview of the field—a little history, where it's at today, and promises of the future; as well as a breakdown of its structure (how it's organized) and a glimpse of some of its many career options.

The "Careers" section includes eight chapters, each describing in detail a specific engineering specialty: biomedical, chemical, electrical and electronics, environmental, mechanical, packaging, plastics, and transportation. The educational requirements for these specialties range from bachelor's degree to Ph.D. These chapters rely heavily on first-hand accounts from real people on the job. They'll tell you what skills you need, what personal qualities you have to have, what the ups and downs of the jobs are. You'll also find out about educational requirements—including specific high school and college classes—advancement possibilities, related jobs, salary ranges, and the future outlook.

The real meat of the book is in the section called "What Can I Do Right Now?" This is where you get busy and DO SOMETHING. The chapter "Get Involved" will clue you in on the obvious volunteer and intern positions and the not-so-obvious summer camps and summer college study, high school engineering programs, and student engineering organizations.

In keeping with the secondary theme of this book (the primary theme, for those of you who still don't get it, is "You can do something now"), the chapter "Do It Yourself" urges you to take charge and start your own programs and activities where none exist—school, community, or the nation. Why not?

While we think the best way to explore engineering is to jump right in and start doing it, there are plenty of other ways to get into the engineering mind-set. "Surf the Web" offers you a short annotated list of engineering Web sites where you can explore everything from job listings (start getting an idea of what employers are looking for now) to educational and certification requirements to on-the-job accounts to practical engineering problems.

"Read a Book" is an annotated bibliography of books (some new, some old) and periodicals. If you're even remotely considering a career in engineering, reading a few books and checking out a few magazines is the easiest thing you can do. Don't stop with our list. Ask your librarian to point you to more engineering materials. Keep reading!

"Ask for Money" is a sampling of engineering scholarships. You need to be familiar with these because you're going to need money for school. You have to actively pursue scholarships; no one is going to come up to you in the hall one day and present you with a check because you're such a wonderful student. Applying for scholarships is work. It takes effort. And it must be done right and often a year in advance of when you need the money.

"Look to the Pros" is the final chapter. It's a list of professional organizations that you can turn to for more information about accredited schools, education requirements, career descriptions, salary information, job listings, scholarships, and much more. Once you become an engineering student, you'll be able to join many of these. Time after time, professionals say that membership and active participation in a professional organization is one of the best ways to network (make valuable contacts) and gain recognition in your field.

High school can be a lot of fun. There are dances and football games; maybe you're in band or play a sport. Great! Maybe you hate school and are just biding your time until you graduate. Too bad. Whoever you are take a minute and try to imagine your life five years from now. Ten years from now. Where will you be? What will you be doing? Whether you realize it or not, how you choose to spend your time now—studying, playing, watching TV, working at a fast food restaurant, hanging out, whatever—will have an impact on your future. Take a look at how you're spending your time now and ask yourself, "Where is this getting me?" If you can't come up with an answer, it's probably "nowhere." The choice is yours. No one is going to take you by the hand and lead you in the "right" direction. It's up to you. It's your life. You can do something about it right now!

Section 1

What Do I Need to Know About

Engineering

?

Earth spaceship lands on Mars?

Yes, thanks to our earthling engineers, *we* are now the aliens. NASA's Pathfinder missions (and mid-1970s Viking missions) to Mars are among this century's most challenging, creative, and awesome engineering feats. Just think of it, we, humankind, envision, create and launch a spacecraft a mind-boggling one hundred million miles through the icy expanse of space to land on a different planet in a predetermined location, and from that craft we deploy a little six-wheel-rover to navigate the rugged terrain via remote control and analyze rocks and soil, take scenic photographs, monitor the weather, and then send back to Earth all the information it has gathered. Wow!

The number of engineering branches that directly and indirectly contributed to these recent Mars missions is impressive: aerospace engineers devised the way to get to and then land on Mars; mechanical engineers designed Pathfinder and the rover; optical engineers were responsible for the amazing images we see of the planet; electrical and computer engineers worked on the complex computer circuitry and electronics involved with space travel; and the list could go on and on. Some of these engineers may work individually, but for the most part, they all work together as a team. The aerospace engineers may have a great idea, but the mechanical and electrical engineers may tell them something like, "No, no, we can't do that." And then the light bulbs go on above their heads and the fun of creating something from nothing begins.

Engineers make things work. Their creative ingenuity impacts our lives and our societies in many ways people never even notice. Not only are they involved in the breathtaking excitement of space travel, but they also have their hands in more down-to-earth, behind-the-scenes projects like designing safe and lasting bridges, designing the hottest new theme park rides, making sure our drinking water is clean and safe, and even figuring out ways to make heavy-traffic areas safer and less congested. In almost every area of the modern world—look around you—the brain power and vision of an engineer is present.

GENERAL INFORMATION

A lot of brainpower goes into engineering, a lot of knowledge, creativity, thoughtfulness, and pure hard work. Humankind has been "engineering," so to speak, since we realized we had opposable thumbs that we could use to handle tools. And from that point on we began our ceaseless quest to make, to build, to create tools and systems that helped us live our lives better. There were a lot of mistakes, but we learned from them and built a foundation of engineering laws and principles.

We can trace the development of civilization to the present day through engineering hallmarks like Stonehenge, the Egyptian pyramids, the ancient cities of Greece, the extensive system of roadways and aqueducts built by the early Romans, Europe's fascination with fortresses and cathedrals, the invention of dams, electricity, the automobile, the airplane, the building of canals and cross-continental railways, nuclear energy, putting a man on the moon—and that's just skimming the surface of the many things engineers are responsible for that built our civilizations and defined how we think of ourselves and our societies.

The rise of the first cities in 3000 BC in Mesopotamia (modern-day Iraq) created a need for engineers, though there was certainly no concept of what an engineer was then. These early "engineers" did not apply scientific principles to their work, but rather they learned by example, from mistakes, and from the urgency of pure need. People relied on "engineers" to address their everyday needs and survival. To protect against enemy attacks, engineers learned to heighten and strengthen building walls through the use of brick (a building material most likely invented by accident). To bring food and water into the city, engineers constructed a system of levees, small canals, and reservoirs.

The first engineer of prominence whose architectural legacy has survived millennia, was the ancient Egyptian builder, Imhotep. He designed and built what is commonly believed to be the first pyramid, the Step Pyramid, around 2650 BC just outside of present-day Cairo, Egypt. Modern engineers marvel at the skills the ancient Egyptians demonstrated in the building of the pyramids. To build a structure as massive and architecturally perfect as a pyramid at a time of very limited building resources (the invention of the wheel was in its infancy—they had nothing like cranes, levels, or any sort of machinery to

Connecting Two Oceans

The building of the Panama Canal was a feat of epic proportions. It was the largest and most expensive engineering project ever undertaken. Costing an estimated $366,650,000, the canal took 33 years to complete from the first shovel in the ground in 1881, to the first boat passing through in 1914.

There were constant setbacks as numerous engineers attempted its completion. The French engineer, Ferdinand de Lesseps (1805-1894), builder of the Suez Canal, was the first to have at this giant task. Almost as soon as work began many of his crew began to die of tropical diseases. Malaria and yellow fever killed thousands. Working conditions in the jungles of Panama were miserable and morale was low. Eight years, $287 million, and 20,000 lives later the whole affair became a scandal and the French called it quits.

This was good news for U.S. President Theodore Roosevelt (1858-1919). He saw the Panama Canal as the key to the United State's rise to that of superpower. He paid the newly created Panama government ten million dollars to resume work and hired engineer John Frank Stevens to oversee the project. Stevens changed the French plan to include a series of locks to carry ships up and down through the canal. In 1907, George Washington Goethals (1858-1928), an army engineer, took over construction of the canal. He overcame endless setbacks and worked on improving labor relations, while still staying true to Stevens' vision. The canal finally opened for business on January 7, 1914, with the unceremonious passing of the Alexandre la Valley.

move the large heavy stones) was a task of incredible ingenuity. The ancient Egyptians used sledges to transport the stones from the distant quarries where they were mined. The main methods they used to move the huge stones (some weighing up to fifty-five tons) higher up as the pyramid grew, was a method called *jacking*, which used wedges and levers to slowly but surely move the stones higher and higher.

The engineers of ancient Greece studied more complex principles of geometry and put them to use in advanced architectural designs. These engineers used five basic machines to help them in their building: the wheel, the pulley, the lever, the wedge, and the screw. More complex building methods and different building materials began to appear with the Greeks. They developed a variety of joints, made use of the column as a load reliever, commonly used post-and-beam construction, and they used the arch, although rarely. The Greek engineers found use for iron, lead, limestone, and marble. Other Greek contributions to engineering include studies on the lever, gearing, the screw, the siphon, and the concepts of buoyancy, and the invention of force pumps, hydraulic pipe organs, and the metal spring.

Whereas the Greeks were the theorists of early engineering, the Romans were the projectors and administrators. They busily set out to construct many great public works, like building roads, bridges, tunnels, aqueducts, and even plumbing for each city home. Rome's major concern for establishing a system of civil engineering was to aide its war machine. Military engineers were responsible for building roads and bridges (to better access future conquests and protect their empire), and baths (to relax the warriors after battle), and of course, for developing a variety of weaponry. The Roman military engineer's greatest responsibility, however, was the fortification of army camps. They built all kinds of fortifications, which included walls of varying thickness, height, and shape to better repel would-be attackers.

Modern engineering's true beginnings are mostly rooted in the seventeenth and eighteenth centuries, where mathematical principles and laws of physics began to be understood and developed. Isaac Newton's (1642–1727) groundbreaking research in mathematics and physics was quickly picked up by engineers and put to practical use. Aside from enlightening the world about gravity, Newton's work in mechanics produced the generalization of the concept of Force, the formulation of the concept of Mass (his First Law), and the principle of Effect and Counter-Effect (his Third Law). Other mathematicians and engineers of the time, enlightened by Newton's findings, went on to make mathematical discoveries that paved the way for the work of future engineers.

It wasn't until the eighteenth century that the first schools of engineering were established. Previously, most young engineers learned their skills by apprenticeships, if they were lucky enough to get one. A French military engineer, Sébastien le Prestre de Vauban (1633–1707), recognized the need to have an actual corps of engineers in the military to study and improve the building of fortifications, bridges, and roads. Shortly thereafter, engineering schools began to appear throughout France, and then the rest of Europe, and finally in the United States. In 1775, the U.S. Continental Congress stated, "That there be one Chief Engineer at the Grand Army . . . [and] that two assistants be employed under him . . . ," and thus began the United States Corps of Engineers. This new need for engineers prompted the beginning of scientific schools at Harvard (1847), Yale (1861), the Massachusetts Institute of Technology (1865), and other engineering schools.

Vauban's plan to educate engineers was largely for military purposes. Armies needed easy ways to get from point A to point B, so roads and bridges had to be built (the Romans, we have seen, did this very well). It was around this time too, that the civil engineer came on to the scene as a separate discipline. When military engineers built roads, bridges, canals, and other public works, they tended to build them only when it served a very specific military purpose. Cities needed better water-supply and sanitation facilities, and local roads that could link smaller communities. As cities grew, they tended to do so in a not-too-orderly fashion, so civil engineers were needed to plan the cities. In 1771, John Smeaton (1724–1792), the first self-proclaimed civil engineer, founded the Society of Civil Engineers with the objective of bringing together like-minded engineers and other men of (financial) power to design and build public works.

With the advent of steam power and the industrial revolution at the end of the eighteenth century, great engineering accomplishments seemed to happen every day—from the steam engine and the large-scale mining of coal, to the cotton gin and new agricultural technology, the world was changing quickly. New processes for manufacturing iron and steel made their use more common in all branches of engineering, which was evidenced by the first suspension bridge, erected by J. Finley in the United States in 1801. This invention made possible the building of bridges in locations where a standard midspan support bridge could not feasibly be built, and greatly improved transportation. A suspension bridge is suspended by cable attached to and extending between supports or towers—a modern-day example is the Golden Gate Bridge in San Francisco. Over the years suspension bridge technology was improved and they became a common sight throughout the world.

//There can be little doubt that in many ways the story of bridge building is the story of civilization. By it we can readily measure an important part of people's progress."—Franklin D. Roosevelt, October 18, 1931

The nineteenth century marked the dawn of electrical engineering. Many scientists at the time, excited over this new, and most likely profitable, phenomenon, electricity, dove headfirst into all its mysteries and intricacies, making many valuable discoveries. Once these scientists laid down the principles of the field, other engineers, inventors, and scientists put these principles to practical applications, like Samuel Morse's (1791–1872) invention of the telegraph in 1837, Alexander Graham Bell's (1847–1922) telephone in 1876, Thomas Edison's (1847–1931) light bulb in 1878, Nikola Tesla's (1856–1943) electric motor in 1888, and many others. By the early twentieth century, much of the infrastructure of modern society as we know it was coming to light.

The first major engineering feat of the twentieth century was Orville (1871–1948) and Wilbur (1867–1912) Wright's first controlled flight of a powered airplane. Aeronautical engineering, as it came to be known, was a dangerous endeavor at its outset, and early engineers labored for a long period of time without significant success. Gliders were the first airplane design to test the skies and provided engineers with valuable information on aerodynamics. The Wright brothers came up with the idea of fixing a motor to their plane for long distance flight and made several failed (though educational) attempts to take off. Finally, in 1903 they succeeded in flying an engine-powered biplane for fifty-nine seconds. Over the next two years they made more adjustments and tests and eventually ended up selling planes to the U.S. military.

//**Engineers . . . are not mere technicians and should not approve or lend their name to any project that does not promise to be beneficent to man and the advancement of civilization."—John Fowler, English civil engineer (1817–98)**

As with previous centuries, it was the military and war that fueled most of our major engineering advancements. The twentieth century has been no different. The U.S. and European militaries recognized immediate potential in airplanes and put its engineers to work on making them safer, more maneuverable, faster, and rugged. By 1914, just in time for the first World War, militaries fitted their planes with radios, navigational equipment, and guns. Planes were also designed to haul and drop bombs. European navies invested heavily into submarine research. The German navy, although a latecomer to the submarine, was the early leader in this technology, and during World War I its U-boats (*untersee boots*) were greatly feared by Allied ships at sea.

Fighter plane and submarine design advanced even further by World War II. Military engineers increased the plane's and sub's speed, range, and general mechanical abilities to make them easier to learn and operate. By 1942 the German Luftwaffe flew the first real jet plane, which reached speeds of over five hundred miles per hour.

One of this century's most profound engineering innovations came in the 1930s with nuclear power. In the race to develop the atomic bomb, the U.S. government spearheaded the Manhattan Project to study, develop, and learn about nuclear power. As nuclear engineers began to realize the awesome peaceful potential of nuclear power, schools were established to study this extremely dangerous new technology and a new branch of engineering was born—nuclear engineering.

In the 1950s, the Cold War brought about fierce technological and military competition between the United States and Russia, necessitating the services of all types of engineers. The exploration of space became another area where engineers sought to demonstrate their country's technological dominance. In 1957, the Soviets launched *Sputnik*, the world's first orbiting satellite. The United States countered with its own satellite, *Explorer I*, in early 1958. In the early 1960s, Russia and America set their sights on the moon and engineers from both countries worked feverishly to be the first to land a man on the moon. And in 1969, the world was awestruck when U.S. astronaut Neil Armstrong stepped from his spacecraft onto the rocky surface of the moon.

Even today engineering marvels still abound, from the skyscrapers in our cities, to the artificial heart, to the tiny computer processors in many of the electronics we buy. The rapid speed at which computer technology is progressing is unprecedented. Big news one month in the computer industry is almost ancient history one, two, three months down the road. Modern society is becoming increasingly automated, as human power is replaced by the loyal, never-complaining robot.

According to the 1828 charter of the British Institution of Civil Engineers, engineering is "the art of directing the great sources of power in nature for the use and convenience of man." Most engineers will hold true to this as an accurate definition of their field, with perhaps a few additions. Not much of the philosophy has changed since 1828. What has changed is the scope and method of engineering, which began as an empirical art and has gradually developed into a highly specialized science.

STRUCTURE OF THE INDUSTRY

The engineering industry comprises many fields of study, all employing unique and sometimes similar methods of science to reach practical solutions to problems and questions in all industries. There are five basic areas of study in which engineers in all branches of the industry can specialize: *research, development, application, management,* and *maintenance*. While an engineer may work exclusively in one of these areas, it is more common for their knowledge and duties to overlap. For example, research and development are commonly linked and called R&D. Those working in the application side of a project may often find themselves working on maintenance calls.

THE INDUSTRY, CONTINUED

All engineering projects begin with *research*. The type and scope of research will be different in each branch of engineering, but generally, once a goal has been set by an industry or institution and brought to the engineers, they set out to find possible ways to solve it. This could entail visiting a manufacturing plant to get a first-hand understanding of an industrial problem. It could entail testing different materials for a specific space application. It could also require the engineer to struggle over mathematical or computer applications, as well as spend long hours in a laboratory mixing chemicals. The scope of research is limitless. Some research engineers work independently, seeking to find new principles and processes for a specific branch of engineering.

"From the point of view of modern science, design is nothing, but from the point of view of engineering, design is everything. It represents the purposive adaptation of means to reach a preconceived end, the very essence of engineering."
—Edwin T. Layton, Jr.

Closely tied to research is *development*. Development entails applying the results of a research project to a specific function. Since there may be more than one way of doing this, a development team must perform tests and studies to find the best way. A research engineer may have come up with several useable materials for a part on an airplane, but it is up to the development team to figure out which of those materials works best in the big picture. For example, a certain material might be chosen for an exterior part on a plane. It works fine, it's durable and weather resistant; however, a development team might discover the material reacts negatively to the exhaust of the engines, or that at certain altitudes and speeds it reacts differently than when on the ground. With these observations made, the development team will reject the material and look for one that's more stable under these types of stresses.

With the research and development complete, the real fun begins—project *application*. Engineers use the data from the R&D studies and apply them to the design and production of materials, machines, methods, or to whatever the ultimate goal is. For example, a team of civil engineers, after finding the best materials and location for building a bridge in Alaska, will then set out to design and build the bridge. A team of industrial engineers, having studied an outdated method of production at a manufacturing plant and having researched newer methods specific to that type of manufacturing, will begin implementing the new method and train plant employees.

A *management* engineer, who earlier may have been part of the R&D or application team, will be responsible for keeping the developed idea working. They study their work as it was intended to function—whether that be machinery, a drawbridge, or a carving knife—and look for ways to improve on it in the future. If a team of electrical engineers designs a new system of electronic circuits for a home entertainment system, they may find that once everything is in place and the system has been manufactured that there are certain minor aspects that can be improved. While the development team is supposed to have found these possible flaws before production, it is almost impossible to catch every possible little glitch. Engineers consider this learning from experience and will improve the design in the future.

Maintenance, the final stage in an engineering project, is concerned with the project's upkeep. A team of mechanical engineers, for example, may have designed a machine to package coffee in tin containers. The engineers visit the plant, set it up, and go home with everyone satisfied. Two months later, a maintenance engineer, who was likely involved in the application process, is called back to the plant to fix a problem in the system—the coffee keeps overflowing, causing waste. These sorts of maintenance calls are fairly common, as it takes a while with any new project to work out the bugs. Again they learn from experience, and similar systems in the future will probably not have the same problem.

Engineers in a particular industry, say in nuclear engineering, will of course, be trained in nuclear engineering, but they will also have to understand many basic to complex principles of other branches of engineering. Depending on their specific studies, the nuclear engineers may need understanding of environmental, chemical, aerospace, mechanical, electrical, industrial, materials, naval, or computer engineering.

Professional Engineer, or PE, is a special title like Ph.D., which indicates that the engineer has completed education and experience requirements and passed tough exams. PEs are under legal responsibility for their engineering work and are bound by an ethics code to protect the public health and safety.

THE INDUSTRY, CONTINUED

PEs have graduated from an accredited engineering program, have had at least four years experience under a licensed PE, and they have passed the Principles and Practice of Engineering (PPE) exam.

CAREERS

If you want a career in engineering you'll have to specialize. However, the engineer specializing in any field is usually required to have basic knowledge of other engineering fields, since most problems engineers face are complex and interrelated. There are over twenty-five engineering specialties recognized by professional societies, and depending on your particular aptitude and interests—whether you're a nuts-and-bolts type person or a computer jock—the doors to a career in engineering are open. The following paragraphs describe some of the most prominent engineering careers.

Biomedical engineers are highly trained scientists who apply engineering and life science principles to problems in medical research and health care. They design health care instruments and devices or apply engineering principles to the study of human systems.

Chemical engineers use their skills and understanding of chemistry, physics, mathematics, and other engineering principles in the design of equipment and manufacturing processes for the production of chemicals and other goods made from chemical processes. They help develop products such as plastic, metal, gasoline, detergents, pharmaceuticals, and foodstuffs. They develop or improve safe, environmentally sound processes while determining the least costly production method for the product.

Electrical and electronics engineers are concerned with developing practical applications of electricity in all its forms. Electrical engineers work with "heavy current" electricity in developing equipment and processes that produce and distribute electricity, such as systems that generate high-power electricity. Electronics engineers work with "light current" electricity in developing virtually anything that uses electricity, from a computer to a camera to a satellite.

Mechanical engineers design, test, build, and maintain all kinds of mechanical devices, components, engines, and systems. One of their chief concerns is the production, transmission, and use of power.

Packaging engineers design the packaging for consumer goods such as food, electronics, medical products, toys, appliances, clothing, and many more. Packages are designed to protect products, provide benefit to consumers, conserve natural resources, and minimize waste through recycling.

Plastics engineers create, design, and test, polymeric materials to manufacture useful end products, from plastic automobile parts to biodegradable polymers for a packaging company to plastic fibers for clothing.

Aeronautical and aerospace engineers (the two branches are closely allied) deal with the development, manufacturing, maintenance and testing of all types of aircraft. Aeronautical engineers must have thorough knowledge of aerodynamics as well as navigation systems and structural design. As an aerospace engineer you may study celestial mechanics (how objects act in space), or you may be concerned with fluid mechanics or structural design. Whatever your specialty, you will have to be an ace math student, as complex mathematics are the norm in this field.

Fast Facts

Young engineers just entering the field with a bachelor's degree can earn as much as 75 to 100 percent more per year than those in nonengineering fields with a bachelor's degree.

Civil engineering is one of the broadest fields in the industry. *Civil engineers* are responsible for the lay of the land. They design roads (including highways, skyways, and small farm roads), bridges, tunnels, harbors, railroads, airports, water supply systems, and many other construction projects in your community. Wherever there is a large construction project under way, a civil engineer is usually in charge or behind the scenes planning and troubleshooting. When the work is done, the civil engineer is responsible for maintenance and upkeep strategies to ensure public safety.

Transportation engineers are civil engineering specialists who plan, design, and operate all methods, structures, and systems that transport people and goods in a safe, convenient, rapid, and environmentally responsible manner. Typical projects they work on are streets, highways, tollways, airports, transit systems, railroads, and harbors.

Industrial engineers plan and implement strategies for the most efficient means of industrial production. They study how the balance of machinery, labor, and raw materials is most effectively and safely put to use in a production process. They work at large industrial manufacturing plants, medical centers, and other large complexes. As an industrial engineer you will have to have a keen sense of economics, as saving money is often a primary objective.

Everything has to be built with something, and that's where the *materials engineer* comes in. Materials engineers develop and test different materials for specific products, from the skin of the Space Shuttle to the soles of your shoes. They may develop more environmentally safe materials or work on

CAREERS, CONTINUED

methods of disposing hazardous materials safely. A materials engineer studies a product's design, and based on the various functions of the product, suggests the most appropriate materials, whether they be metals, alloys, plastics, composites, fibers, or other materials.

Nuclear engineers are concerned with the design, construction, and safety of nuclear reactors, radioactive waste sites, and devices using nuclear energy. They study ways to apply radiation in the diagnosis and treatment of diseases. In addition to power production and health applications, nuclear engineers find work in space exploration, agriculture, environmental concerns, and transportation.

The *petroleum engineer* seeks the best location and method of tapping (drilling) an oil or natural gas reservoir to extract as much from the reservoir as possible. They are also responsible for finding the most economic means of developing oil or gas sites. Some petroleum engineers may specialize in geological engineering, which is concerned with finding accessible oil and gas reservoirs. Petroleum engineers are skilled in geology, physics, and the engineering sciences.

Safety engineers are concerned with preventing accidents. They work in many different branches of engineering, from mechanical to transportation engineering. As a safety engineer you may study *ergonomics* to develop human-friendly machines that reduce repetitive motion disorders, or you may plan fire evacuation plans for a large factory or office complex. Safety engineers are found at construction sites, environmental hazard spots, insurance companies, transportation offices, and similar workplaces.

Fast Facts

Good news for the nuclear engineer: By the year 2000, an estimated one-third of all electric power generated worldwide will come from nuclear power plants.

EMPLOYMENT OPPORTUNITIES

Since engineering is such a broad field, there are literally thousands of different places you can find employment, depending on your specialty. Engineers are needed in virtually every field. Whether you want to work for a small company or a large firm, indoors or outdoors, 9 to 5 or the graveyard shift, there are engineering careers available for you. You can find your engineering career at a desk behind a computer with regular hours not far from your home; or you may find your engineering career in the workshop with a socket set in one hand, a computer-aided calibrator in the other, and a little grease under your nails; or you may find your engineering career in the dank sewers of a major city as you plan a more effective method of waste management.

As a mechanical, automobile, chemical, industrial, plastics, or robotics engineer you may find employment with one of the Big Three U.S. automobile makers, as well as any of the thousands of private manufacturing companies. As an aeronautical or aerospace engineer you will almost certainly try to land a job at NASA or a major commercial firm like McDonnell Douglas (which recently merged with Boeing). Civil engineers can find jobs with local and city governments, with construction firms, with the military or federal government, and even large corporations. The petroleum and chemical engineer can seek jobs, naturally, with petroleum and gas companies like Amoco, Texaco, and Exxon, or they can look for jobs with major chemical companies like Dow Chemical, Eastman Chemical Company, or E. I. DuPont. Environmental and biological engineers will have no problem finding employers at the Environmental Protection Agency, industry, or consulting firms. Companies producing high-tech equipment for commercial and industrial use look for skilled electrical engineers, as well as software, mechanical, materials, and plastics engineers.

Engineers are also employed by local, state, and the federal government—178,000 in 1996, according to the U.S. Department of Labor. Some federal employers of engineers include the National Aeronautics and Space Administration, and the Departments of Agriculture, Defense, Energy, Interior, and Transportation.

Other possibilities for engineers can be found in academia as instructors or researchers.

INDUSTRY OUTLOOK

We live in a rapidly changing, high-tech world, where the major riddles of one year are solved the next year. Who's behind this? Well, largely it's engineers. Engineers wield tremendous influence on our lives, cultures, and societies. As the world changes with the latest technologies and as we grow accustomed to the luxuries they offer, there will be an increased demand for highly skilled, innovative professionals to continue the trend. These advancing technologies will put pressure on industry to improve and update their current systems— whether they be for production facilities or corporate offices—if they are to stay competitive. Some branches of engineering are certain to remain strong for the long haul, like civil, mechanical, electrical, computer, and safety.

OUTLOOK, CONTINUED

In the past ten years the computer industry has grown at an unprecedented rate. While its volatility scares some people away, the field has a great need of skilled engineers to continue its rapid growth, so job prospects and outlook are strong for the long term. The computer industry is fiercely competitive. Cutbacks and layoffs are common, but those engineers who keep up to date with the current high-tech trends will have little trouble locating another job. The U.S. Bureau of Labor Statistics predicts that job possibilities will grow 103 percent for systems analysts, 109 percent for computer engineers, and 118 percent for all other computer scientists through the year 2006.

Cities and neighborhoods are always changing. Roads and bridges age or wear out and need to be repaired or rebuilt. For this reason alone civil engineers will always be in demand. Traffic in our cities and suburbs continues to be a problem and will require the attention of transportation engineers. As the atmosphere becomes more polluted and begins to negatively effect our lives, measures will be taken to find alternative energy sources. While this may be somewhat bad news to the petroleum engineer, any affect on the industry will be offset by the efforts required of other engineers to find a solution.

As we continue to damage the environment (a tragedy some engineers must take responsibility for), environmental and biological engineers will find themselves busy. Many governmental agencies and big businesses are becoming more sensitive to environmental concerns, especially as complex and stringent regulations pass through Congress. Despite this, recent deficit reduction strategies have meant less governmental money for major projects such as the Superfund cleanup.

Due to recent cuts in defense spending, aerospace engineers have had a tough time lately. Stiff competition among nongovernmental employers has meant that companies work from a tight budget and are not likely to expand production. Currently, there are more qualified job candidates than there are jobs. While aerospace jobs in research and development remain stable, few new jobs will be added.

Careers

?

engineering

Biomedical Engineer

SUMMARY

DEFINITION
Biomedical engineers *apply engineering principles to problems in medical research and health care.*

ALTERNATIVE JOB TITLES
None

SALARY RANGE
$25,000 to $38,000 to $65,000

EDUCATIONAL REQUIREMENTS
Minimum of a bachelor's degree. Advanced degree recommended.

CERTIFICATION OR LICENSING
Voluntary

EMPLOYMENT OUTLOOK
Faster than the average

HIGH SCHOOL SUBJECTS
Biology
Chemistry
Mathematics
Physics

PERSONAL INTERESTS
Building things
Computers
Figuring out how things work
Science

Wrapping up the report he is working on, Mike Vonesh is ready to go home. As he checks his planner to see what he has scheduled for the next day, his phone rings.

"There's a Dr. John Mackey for you on line one," says his secretary. "Do you want to take it?" Mike remembers the name. Dr. Mackey is a surgeon working in Denver. They'd met at a convention the year before. "Put him through. . . . Hello, Dr. Mackey, how are you?" "Mike! I'm glad I caught you. I remember we spoke about blood vessel technology, and I just thought—" There is a long pause. When Dr. Mackey continues, he sounds tired. "I lost a patient today. Aortic aneurysm. He was only sixty, but he had diabetes and previous heart attacks, so we just couldn't operate. Mike, I don't ever want to look a patient like this in the eye again and tell them there is nothing I can do. There has to be a way to treat this without opening them up."

They talk for a while, the doctor who wants to save lives and the engineer who can help him. Mike tells Dr. Mackey about the advisory group of surgeons he works with and asks the doctor to join. Dr. Mackey agrees, and by the time they say goodnight, Mike has a page of notes and a new member of his team. As he leaves the office, his mind is filled with possible solutions and ways to test those ideas. He knows that with the right tools, doctors like John Mackey can save many more lives. Mike's job is to give them those tools.

WHAT DOES A BIOMEDICAL ENGINEER DO?

Biomedical engineers bridge the gap between the mechanical world and the world of flesh and blood. They use their understanding of engineering to help solve problems in health care. They work in cooperation with doctors, technicians, and engineers from a variety of fields to develop and test machines, materials, and techniques that give patients hope for longer, fuller lives.

On the surface, the fields of engineering and medicine seem far apart. Engineers work with steel, fluids, and mathematical principles to create new and better things. Doctors work with the most amazing machine—the human body. Ever since the first artificial limb was used to replace a missing leg, however, engineering and medicine have worked together. Engineers use their knowledge of physical laws to study the operations of the natural world. Better understanding of how systems within the body work can lead to new treatments and tools for doctors. The wonders of modern medicine, from the simplest artificial limbs to artificial hearts, are due in part to the work of biomedical engineers.

There are a number of subfields within biomedical engineering. Some people in these fields work in basic research, exploring theories and broad concepts. Others use that knowledge to build, test, and eventually market products or equipment—such as implants like titanium hip joints or diagnostic machines like the ultrasound machine used to examine organs and tissues without surgery. Engineers also work in adapting these new devices to the medical environment by customizing software and training personnel.

Whatever their role, biomedical engineers work closely with people from many different fields. They need the input and cooperation of doctors and medical scientists, but they also work with marketing and sales departments, foundations that issue grants, and machine shops that build prototypes of new devices. Specialists to some extent, biomedical engineers are also generalists who know something about a variety of disciplines.

Biomedical engineers can work in a number of different environments, depending on their specialty. For example, clinical engineers work mostly in

Lingo to Learn

Bioinstrumentation: *Building machines for the diagnosis and treatment of disease.*

Biomaterials: *Anything that replaces natural tissue. These can be artificial materials or living tissues grown for implantation.*

Biomechanics: *Developing mechanical devices like the artificial hip, heart, and kidney.*

Cellular, tissue, and genetic engineering: *Application of engineering at the cellular and subcellular level to study diseases and design intervention techniques.*

Clinical engineering: *Application of engineering to health care through customizing and maintaining sophisticated medical equipment.*

Systems physiology: *Using engineering principles to understand how living systems operate.*

hospitals, while engineers who have moved into sales or marketing may spend more time traveling.

WHAT IS IT LIKE TO BE A BIOMEDICAL ENGINEER?

Mike Vonesh works in artificial blood vessel product design for W. L. Gore and Associates in Flagstaff, Arizona. Before that, he worked at Northwestern Memorial Hospital in Chicago. His days vary, but there is one constant: "There is a lot of interaction with other specialists," he says. "That's a big part of the job—working with other people." Some of this communication is with other engineers working on related products or doctors seeking (or giving) advice, but not all of it. "You have to be able to talk with all kinds of people, from CEOs to the machinists who help build the models."

"You have to be able to talk with all kinds of people, from CEOs to the machinists who help build the models."

Mike estimates that about 75 percent of his time is spent in "hands-on" work—running tests, performing animal studies, or working on models. The remaining 25 percent is spent in meetings or other administrative duties. "I have a lot of long telephone conversations," he says.

Mike divides his job into four phases. Phase one is problem identification. What exactly is the nature of the problem? Doctors may approach him about a problem, as Dr. Mackey did. Or, Mike's company may identify a situation where they might be able to fill a need. Once the problem is isolated, the second phase, problem solving, begins. "I spend a good percentage of my time in problem-solving sessions," Mike says. "I'm a fairly creative person; engineering is really like an art form in the conceptual phase." Much of this phase is devoted to making sure the solutions that are proposed are feasible. Mathematical models are created, often with the aid of computers. If the theories are sound, the process moves on to the next phase: turning the ideas into reality. Prototypes are built and tested in laboratories. Some of the tests are on animals. The data is collected, modifications are made, and new models built and tested. Once the concepts are tested and proven to work, the final phase begins. There are patient trials, where the new device or technology is tested on humans under carefully supervised conditions. If the company decides that there is a market, and government regulatory agencies like the Food and Drug Administration approve, the new device can go into production. The whole process takes between eighteen months and two years.

These steps can overlap. At Gore and Associates, for example, steps are taken to keep regulatory agencies abreast of developments during each phase of development. "We have a long history of cultivating relationships with regulators," explains Mike. That way, problems can be identified early, before they become serious. Engineers working in basic research would not be involved in product design, as Mike is, but the essential goals of their work are the same as his. Biomedical engineers are problem solvers.

HAVE I GOT WHAT IT TAKES TO BE A BIOMEDICAL ENGINEER?

Engineering and medicine are among the most demanding fields of study, attracting students of the highest caliber. "If you aren't dedicated, you won't make it through," Mike says. Those interested in biomedical engineering must be good students prepared to study hard. Interests in math and science are important. You must have good problem-solving skills and an inquisitive mind. Problem solving often requires a new approach, so the ability to think creatively is critical. As Mike puts it, "The big hurdle is translating ideas into reality. You have to be able to think 'outside the box.' I think tinkering, inventor-type skills are very important."

Biomedical engineers are often links between very different areas of expertise. For this reason, it is necessary to learn about other fields, such as electrical, material, and chemical engineering. Because they spend so much time talking with other professionals, they need to be excellent communicators. Technical skills aside, Mike feels that the key element to being a biomedical engineer is compassion. "You can see a real difference between biomedical engineers and other types of engineers," he says. "They have the philosophy that they want to give something back."

As with any job, there are difficulties. Working with a number of different fields means there is always more to learn. "I wish I

To be a successful biomedical engineer, you should:

Have good problem-solving skills

Be a good communicator

Be able to get along with many types of people

Be a compassionate person

Have an aptitude for math and science

Have an inquisitive mind

had more knowledge," Mike admits. "It's frustrating, but you can never know everything about every field." Biomedical engineers face challenges other engineers do not. Instead of dealing strictly with materials that have known properties, engineers like Mike must grapple with biological systems that are never the same. "Engineers tend to be perfectionists. It's difficult to have optimal solutions. You have to make compromises."

HOW DO I BECOME A BIOMEDICAL ENGINEER?

EDUCATION

High School

The course of study for biomedical engineering is very demanding. High school students should take as much math and science as possible, including trigonometry, calculus, biology, physics, and chemistry. Communications classes are good, as are problem-solving classes (like logic). Those living near a facility that does biomedical work may wish to arrange a tour or talk to professionals to see if biomedical engineering is of interest.

Some Products of the Biomedical Industry

Artificial hearts and kidneys

Artificial joints

Automated medicine delivery systems

Blood oxygenators

Cardiac pacemakers

Defibrillators

Laser systems used in surgery

Medical imaging systems (MRIs, ultrasound, etc.)

Sensors that analyze blood chemistry

Postsecondary Training

The minimum degree required for working in biomedical engineering is a bachelor's degree, but usually an advanced degree is required. Mike's experience may be instructive. He graduated second in his class from the University of Illinois, one of the best engineering schools in the country. He was unable to find a job, but he still wanted to be in the field. He decided to go to the University of Arizona for more schooling. After spending two years there, he got a job at Northwestern. He earned his Ph.D. while working full-time. Most biomedical engineers have a bachelor's degree in biomedical engineering, or a related field in the sciences, and a Ph.D. in a biomedical engineering specialty.

Undergraduate study is roughly divided into halves. The first two years are devoted to theoretical subjects like abstract physics and differential equations in addition to the core curriculum most undergraduates take. The third and fourth years are spent with more applied science. "I didn't like the theory very much," Mike admits. "The hands-on classes were more interesting to me, and I got better grades."

There are a number of excellent graduate-level programs. The University of Arizona, where Mike went for two years, was involved in the development of the first artificial heart implanted in a human, the Jarvic 7. In addition to classroom work, students in graduate programs work on research projects headed by permanent faculty. Mike was able to work on projects in partnership with some of the best-known companies in the field, like Boston Scientific and Baxter. In the past, basic research was done at universities and product development was done by private companies. "I think there are more combined efforts now," Mike says. "A lot of the cutting-edge research is being done by small companies that may have money, but lack the kind of intellectual resources you find in universities."

CERTIFICATION OR LICENSING

Mike has a professional engineering certification through the State Board of Technical Registration. This is a voluntary step most engineers do not take. Candidates must have a bachelor's degree in engineering, after which they serve a five-year residency before taking a comprehensive exam.

INTERNSHIPS AND VOLUNTEERSHIPS

It stands to reason that the more "real world" research experience people have, the better their chances of finding a job after school. Internships are one way of getting that type of marketable experience, as well as a way to network with possible employers. "Gore sponsors about a dozen interns every summer, and other major companies have similar programs," Mike says.

WHO WILL HIRE ME?

Biomedical engineers work at universities, hospitals, in government, and in private industry. A great deal depends on what type of engineering a person does. Rehabilitative and clinical engineers may find more opportunities in hospitals, for example. Research and product development are done at private companies like Boston Scientific, Johnson and Johnson, and Baxter. Universities hire professional engineers to work in their labs alongside professors and graduate students. Government funding that in the past could be counted on to support basic research at universities has been one of the victims of budget cuts in recent years. Some of this loss has been replaced by private industry through cooperative partnerships. Federal and state regulatory agencies must approve new devices and procedures before they can be marketed. On the federal level, the Food and Drug Administration is the primary employer of biomedical engineers.

WHERE CAN I GO FROM HERE?

There are a number of career paths for biomedical engineers. Mike plans to stay with product development, at least for now. "I may decide to leave at some point," he says. "There's a lot of mentoring in this profession, and I feel like that's where I am now, where I want to do that myself." Biomedical engineers can continue researching, move into management, or get into government service. The interdisciplinary nature of biomedical engineering means that those in the field often have the background and experience to move easily into other dimensions of business like marketing and sales. Established professionals can become consultants. As new technologies are developed, there are opportunities for entrepreneurs to go into business for themselves. They might not all be as successful as Bill Gore (Gore and Associates employs thousands of workers in four divisions), but bioengineering is a field where there are new developments all the time.

WHAT ARE SOME RELATED JOBS?

The U.S. Department of Labor classifies biomedical engineer under the heading Life Sciences. Also under this heading are people who work with animal care and research, plant biology, agriculture, and food research. Some of these include: pathologist, laboratory animal care veterinarian, anatomist, animal breeder, range manager, dairy scientist, geneticist, pharmacologist, botanist, plant breeder, soil scientist, aquatic biologist, physiologist, food chemist, and research dietitian.

Related Jobs

Anatomists

Animal breeders

Aquatic biologists

Botanists

Dairy scientists

Food chemists

Geneticists

Pathologists

Pharmacologists

Physiologists

Plant breeders

Range managers

Research dietitians

Soil scientists

WHAT ARE THE SALARY RANGES?

Salaries vary greatly depending on education, experience, and place of employment. Federal employees with bachelor's degrees can expect to start in the mid-$20,000 range. In private industry, salaries are somewhat higher. According to a study done in the mid-1990s, biomedical engineers with bachelor's degrees were earning $29,239. The mean income was approximately $38,000. College professors and Ph.D.s in private industry can earn $65,000 per year and more, depending on the number of years of experience they have. Those with established research credentials can earn substantially more as consultants to business and government.

WHAT IS THE JOB OUTLOOK?

Job growth is expected to be stronger than average as the population ages and the health care industry continues to grow. Generally, older people require more tests and procedures than younger people do. And while the industry is changing rapidly as managed care assumes a more dominant role, the search for better (and cheaper) ways of treating patients will continue to drive bioengineering research and development. MRI, ultrasound, CAT scans, PET scans, artificial hearts, kidney dialysis, tissue replacement technology, artificial joints and limbs, and dozens of other advances in medicine are the products of bioengineering. Biomedical engineers and the professors to train them will be needed to produce the tools for the next generation of treatments.

Chemical Engineer

SUMMARY

DEFINITION
Chemical engineers *use their skills and understanding of chemistry, physics, mathematics, and other engineering principles in the design of equipment and manufacturing processes for the production of chemicals and other goods made from a chemical process.*

ALTERNATIVE JOB TITLES
None

SALARY RANGE
$39,000 to $53,000 to $110,000

EDUCATION REQUIREMENTS
Bachelor's degree

CERTIFICATION OR LICENSING
Voluntary, but highly encouraged

EMPLOYMENT OUTLOOK
About as fast as the average

HIGH SCHOOL SUBJECTS
Biology
Chemistry
Computer science
Mathematics

PERSONAL INTERESTS
Computers
Figuring out how things work
Reading/Books
Science

On a flight bound for Germany, Rebecca Rosenberg reviews her equipment specifications book for a project she designed. She has her laptop out and runs a few more computer simulations to make certain she has not overlooked anything. As she expected, the tests check out. The two-column reformat splitter she designed should produce the different cuts of hydrocarbons her client in the United Arab Emirates needs to make gasoline.

The next day Rebecca meets her clients in a downtown Frankfurt office. She lays out the piping and instrument diagram and points out some of the important details. Her clients raise an objection over a specification that is out of the ordinary. Rebecca is prepared for this question, and to ease their concern she gathers her clients around a computer and runs a simulation process of her design. She holds back a smile as she sees they are impressed. The meetings go on all week and Rebecca fields questions on all different aspects of the project. With everyone satisfied she gets the green light from her clients and phones her Chicago office with the good news. Production will begin immediately.

WHAT DOES A CHEMICAL ENGINEER DO?

In the most simple of terms, chemical engineers study ways to convert raw materials into finished products. Specifically, chemical engineers invent new types of materials, synthesize new or existing materials, transform combinations of elements of matter, and then develop the process by which these same chemical changes can occur safely, efficiently, and on a large scale.

Whether they work for a plastics manufacturer, a processed-foods corporation, or a drug development company, chemical engineers are concerned with applying their scientific and technical knowledge in innovative ways to solve problems. The end result is always the creation of a useful material; be it a garbage bag that rapidly decomposes, a fat-free dessert, or a cure for the common cold.

Finding the solution isn't valuable if that solution can't be applied in practical situations. For example, Dr. Selman Waksman (1888–1973) discovered in the late 1930s that certain chemical combinations were successful at destroying harmful bacteria. Calling his chemical warriors, "antibiotics," he envisioned saving millions of lives. Unfortunately, the chemical compounds Dr. Waksman had discovered could only be produced in small quantities in his laboratory. Of what use were they to the average man if they were too expensive to mass-produce and distribute? Thankfully, Dr. Waksman enlisted the help of chemical engineers who first designed a method of reproducing the chemicals en masse by mutating their structure, and then developed special methods for "brewing" the chemicals in huge, oversized tanks. Their efforts allowed these new wonder drugs to be produced in a cost-effective manner, guaranteeing their impact on human lives around the globe.

Today, chemical engineers work on the large-scale preparation of substances in production plants. Their goal is to find safe, environmentally sound processes; to make the product in a commercial quantity; to determine the least costly method of production; and to formulate the material for easy use and safe, economic transport.

Lingo to Learn

First law: In all processes energy is simply converted from one form to another, or transferred from one system to another.

Heat exchanger: A device that allows the heat from a hot fluid to be transferred to a cooler fluid without the two fluids coming into contact.

Lixiviation: The process of separating soluble components from a mixture by washing them out with water.

Second law: Heat cannot pass from a cooler to a hotter body without some other process occurring.

Thermochemistry: The study of how heat affects chemical substances, especially after a chemical reaction.

Thermodynamics: The study of heat and other forms of energy.

Transmutation: The transformation of one element into another by radioactive decay or by the bombardment of the nuclei with particles.

Chemical Engineering's Greatest Hits:

1. **Air.** *Chemical engineers designed ways to process and use the main components of the air we breathe—oxygen and nitrogen—in numerous important applications, such as steel making, welding, the production of semiconductors, and deep freezing food.*

2. **Atoms.** *The world became a different place after chemical engineers split the atom and isolated its isotopes. Medicine, power generation, biology, and metallurgy have all advanced considerably since this discovery.*

3. **Crude oil.** *Crude oil (unrefined petroleum) is used in more ways than you might think. It is the building block for such things as synthetic fibers and plastics, and can be used in the making of soaps, cosmetics, shower curtains, and shampoos.*

4. **Drugs.** *Antibiotics were rare and extremely expensive in the early part of this century. Chemical engineers did not invent antibiotics, but they did invent ways to make their production larger and cheaper so that the general public could afford them.*

5. **Environment.** *While they may not have been as responsible in the past, today's chemical engineer works hard to clean up past problems as well as develop new manufacturing processes that are environmentally friendly.*

6. **Food.** *World crop production is much higher than it was fifty years ago, thanks to the work of chemical engineers. They have produced fertilizers and pesticides that made crops healthier and more resistant to disease.*

7. **Plastics.** *Plastics are everywhere. Try to imagine life without them. Chemical engineers have played a large role in the creation of plastics, from the nylon used in clothing to the polystyrene commonly used today in things like milk cartons.*

For most chemical engineers, the steps outlined above translate into what is known as *process engineering*. Process engineering involves creating new processes to meet the specific needs of a given product, including selecting or designing the proper equipment, supervising the construction of a plant and/or the production system, and overseeing tests, or "pilot runs." In addition, the chemical engineer is often called upon to monitor the day-to-day operations of the production plant itself, or one of the areas within the plant. The chemical engineer's work isn't finished once the system or plant is in place; on the contrary, the chemical engineer constantly tries to improve the product, the system, the production, or any combination of these aspects of the process, working to make them safer, more efficient, more cost-effective, or more friendly to the environment.

Among the typical industries in which chemical engineers work are chemical, fuel, environment, aerospace, plastics, pharmaceutical, and food. In the most common field of these industries, manufacturing, chemical engineers are busy finding out ways to convert raw materials into products. These products can be as specialized as a liquid coating for a part on a rocket engine, or as common as a new longer-lasting sole for your shoes.

Depending on what industry they work in, chemical engineers' duties vary. Chemical engineers working in *environmental control*, for example, develop strategies to reduce or alleviate pollution at the source, as well as treat those wastes produced by their plant which cannot be eradicated. *Safety officers* are responsible for designing and maintaining

WHAT DOES A CHEMICAL ENGINEER DO?, CONTINUED

plants and processes that are safer for both workers and the community. Chemical engineers working in the specialty of *biomedicine* may work with physicians to develop new methods of tracking the chemical processes taking place within the human body, or they may develop artificial organs that temporarily replace real organs until donor organs are found. In general, chemical engineers are employed in the research and design, development, production, and management areas of companies, but many of their duties may require that they work in more than one, or all, of these areas. In fact, engineers frequently work closely with others to solve problems unique to their area of expertise.

No matter where a student's interests lie, he or she can find a rewarding career in chemical engineering. Even within a specific area, there are many applications, and, therefore, careers. For example, chemical engineers work in food production to design better ways of producing particular types of food *and* they work to develop environmentally safe pesticides and fertilizers. Chemical engineers also work as consultants to the government, computer systems designers, environmental lawyers, patent lawyers, and even brewers of specialty beers. In short, the field is only as limited as the imaginations of those who work in it.

WHAT IS IT LIKE TO BE A CHEMICAL ENGINEER?

Out of college for two years with a bachelor's degree in chemical engineering and English, Rebecca Rosenberg enjoys her challenging job as a chemical engineer for a petroleum company. As part of the design team, she designs an assortment of processes to make a variety of petroleum products but mainly gasoline. Rebecca's love for challenges led her to engineering, and while in college her skills in the sciences (mathematics and chemistry) led her to study chemical engineering. "I just have a natural inclination in this field," Rebecca explains. "Chemistry and math were always my best subjects, and you put the two together and you have a chemical engineer."

Every day Rebecca is faced with challenges. "I really love the rush I get when I solve a difficult problem," she says. "These problems stay with me all day and seem to work themselves out in my head at the oddest times, like when I'm doing the dishes or at the grocery store. I have to carry a little notebook with me all the time so I can write down my ideas."

Working for the design team, Rebecca spends about 20 percent of her day fielding questions from customers on past designs. "Interspersed throughout the day I get faxes, phone calls, emails, and questions from other members of my team about designs which are already out the door," she says. "This can be challenging since I might be in the middle of working on something new, and I'm not thinking about a project I worked on a year ago. We keep all the books for all the projects we have, and I usually have to consult those to answer their questions." These inquires from customers always take priority over whatever she is currently working on.

*II*We have clients all over the world. I've been to Europe, Southeast Asia, and throughout the U.S. on business for my job."

Although it depends on which phase she's in for the project she's working on, Rebecca usually spends the other 80 percent of her day sitting at her desk working on the design on her computer. "In the initial phase of a project," Rebecca explains, "most of what you do is computer simulation." In this process the modeling of the project is done and computer simulations performed to see if the design will do what it's intended to do. It is an extremely complex process and requires very precise detailed work.

Other phases Rebecca might be involved in are the *proposal phase,* where the basic logistics of a plan are laid out, from the scope and size of the project to its cost; the *design basis phase,* where the guarantees of a project are established, including a detailed outline of how the design will be performed, and the schedule for the project; the *process engineering phase,* where the detailed processes of the design are worked out on a computer through mathematical calculations, drawings, and computer simulations; and the last phase, the *project engineering phase,* where the results for the process engineering phase are used to design each piece of equipment. They produce an equipment specifications book filled with detailed designs, instructions, and charts. This is basically a how-to book, which can be handed to any contractor to show how to make the equipment.

Rebecca travels frequently for her job. "We have clients all over the world. I've been to Europe, Southeast Asia, and throughout the U.S. on business for my job," she says. She often must visit sites under construction, or laboratories where potentially dangerous tests are being run. At these sites safety is Rebecca's biggest concern. "Safety regulations vary in different countries. I've been at some pretty scary sites where I just want to get out of there as fast as possible. This is certainly the biggest con of my job, having to be at a pilot plant somewhere where safety regulations just aren't what they should be."

HAVE I GOT WHAT IT TAKES TO BE A CHEMICAL ENGINEER?

Albert Einstein once said, "No amount of experimentation can ever prove me right; a single experiment can prove me wrong." Like all scientists, chemical engineers are trained to pay attention to every aspect of their work, from the largest, most obvious elements, to the most minute details of individual experiments. "In our work to solve problems," Rebecca explains, "if we don't pay attention to every detail, every decimal point, we can really suffer some serious setbacks." In addition to the technical importance of recognizing how small, seemingly minor aspects can have drastic effects on the overall success of a project, there is also a compelling, practical reason for paying close attention to details; chemical engineers often work with very dangerous chemical processes, so any error leading to a malfunction can quite possibly represent life-threatening consequences.

It should go without saying that chemical engineers must practice scientific methods, and among the personal qualities intrinsic to exercising those methods are an inquisitive nature and excellent analytical and problem-solving skills. "Chemical engineers are driven by challenges—we are often forced to be innovative," Rebecca says. "We are constantly asking questions, 'What if we did this instead of this?'" Patience, in these situations, is more of a necessity than a virtue.

In order to solve problems, chemical engineers must first identify and then prioritize problems (as problems often come in batches), determining how each fits into the larger picture that is their project. Every project has a host of issues that threaten a successful resolution. Chemical engineers, like other scientists, must learn how to differentiate and rank these issues in order of importance.

Not to be underestimated are the verbal and written skills which are necessary in any field, but especially in a field dependent on

To be a successful chemical engineer, you should:

Be a stickler for detail and have strong problem-solving skills

Have good communication skills, both written and verbal

Be inquisitive

Have an aptitude for math and science

Have patience in order to solve problems that seem unsolvable

Be willing to travel

communicating complicated data, experiments, and systems. Because Rebecca often deals directly with the clients she's designing a project for, she needs to be able to communicate clearly. Each day she fields phone calls, writes proposals, goes to meetings, as well as works closely with her colleagues. Her ability to quickly and clearly state her ideas, opinions, and supporting data is vital. Rebecca credits her good communication skills to her studies of English while in college. "Chemical engineering isn't my life," she explains. "I read novels all the time and I'm a pretty active letter writer."

Because projects are always coming into Rebecca's office, she always has new designs to work on. "One of the things I love about my job is that it doesn't get old. There's always some new challenge around the corner. You won't get bored being a chemical engineer."

Finally, travel can be a large part of the chemical engineer's role in implementing a successful product, design, or system. While not all engineers go out to sites, laboratories, and pilot plants, many of them do. This may mean traveling down the street, to a neighboring suburb, out of state, or perhaps, even out of the country. Rebecca estimates that she travels on business at least one week out of every month, or roughly twelve weeks a year.

How Do I Become a Chemical Engineer?

EDUCATION

High School

In high school Rebecca still wasn't sure what she would be doing in college and took both science and English classes. "One thing I would recommend to anyone thinking about a career in any engineering field is to get as much experience with computers as possible," Rebecca says.

Potential chemical engineers should consider taking as many science, mathematics, and computer science courses as possible. Typical classes that will help provide a solid base for further, advanced work in this field are chemistry, math (including algebra, geometry, calculus, and trigonometry), physics, and computer sciences. Courses in English and the humanities are also crucial in helping to develop strong communication and interpersonal skills. Foreign languages are another wise investment; in today's global marketplace, the knowledge and familiarity with another culture's customs and language can not only mean the difference between success and failure, but it can also help to set you apart from the crowd when it's time for promotions or to search for a new job. Rebecca, for example, studied German in high school and finds it very useful on her business trips to Germany.

Postsecondary Training

Students dreaming of a career in chemical engineering should be aware that the minimum requirement for entry-level positions as chemical engineers is a bachelor's degree in chemical engineering. Before they spend the time and money applying to various schools, potential students should make certain that the college or university they plan to attend has been approved by the Accreditation Board for Engineering and Technology and the American Institute of Chemical Engineers; there's no sense working toward a degree in an unaccredited program. Currently, 150 colleges and universities offer accredited programs. Most programs last four years, although some may last longer and require a commitment to graduate study.

In the first two years of typical chemical engineering programs, students study chemistry, physics, mathematics, computers, and some basic engineering courses like materials science and fluid mechanics. "These classes are so important because they are the basis of all that we do," Rebecca says. Similar material is covered in high school, but these college courses present students with a much more advanced view.

In the third and fourth years students take classes in advanced chemistry, heat balance, materials balance, reactor design, mass transfer, heat transfer, thermodynamics, and chemical process economics, to name but a few. "This is when you find out if chemical engineering is for you," Rebecca explains. "It's really hard work, but it's also fascinating and can be fun."

Most programs require students to take some classes in the humanities, as well as a technical writing course, since communications and writing are an important part of the job. Rebecca also studied English literature and insists it has helped her immeasurably in her communication skills. "You'd be surprised at how poorly some chemical engineers write," she says. "I'm writing documents and proposals all the time, and I need to be able to express my ideas clearly."

Although bachelor's degree graduates face a wide variety of career options, those who obtain higher degrees undeniably enter their careers at a higher, more specialized level than do their colleagues with only bachelor's degrees. Many positions in research and teaching are open only to MS and Ph.D. candidates and, of course, the salaries for those with advanced degrees are considerably higher.

CERTIFICATION OR LICENSING

Certification in chemical engineering is called licensing. A licensed engineer is called a PE (Professional Engineer), just as a medical doctor is called an M.D. While licensing is voluntary it's quickly becoming mandatory in many areas of this field. Employers want to be able to advertise their companies as containing only Professional Engineers. Not only does it reflect well upon them, but it also means they know they've hired quality engineers, or that they will soon have quality engineers. Licensing also assures employers that they meet state legal regulations.

Requirements vary from state to state but generally it takes about four to five years to become a licensed PE. Students can begin the process while still in college by taking the Fundamentals of Engineering (FE) exam. The exam lasts eight hours and covers everything from chemistry, physics, and mathematics to more advanced knowledge of engineering sciences. Rebecca took the test in her senior year of college and passed it but she knows some people who didn't. "It's not an easy exam," she says. "You really have to prepare for it."

The next step to acquiring the designation of professional engineer is obtaining four years of progressive engineering experience. Some states require you to obtain your experience under the supervision of a PE. Once your four years are up you can take the Principles and Practice of Engineering (PPE) exam specific to chemical engineers (there's one for almost every other branch of engineering, too). If you pass the rigorous exam you can officially call yourself a Professional Engineer.

INTERNSHIPS AND VOLUNTEERSHIPS

Many schools require that their degree students fulfill an internship. Intended to provide students with hands-on work experience, the internship experience also grants students the opportunity to network and gain a deeper insight into how the industry works. If an internship is required, it usually lasts from four to twelve months, or roughly somewhere between one and two semesters. Schools are usually instrumental in locating internships, but placing cold calls or writing query letters to companies you have already carefully researched can also be an effective way of locating a quality internship. Internships required by the school are usually non-paying; students are often compensated in the form of credit toward semester hours.

WHO WILL HIRE ME?

Unlike many recent college grads, Rebecca's job hunt was easy. Actually, it was not even a hunt. "Companies were knocking on my door," she says. "Many of my friends in the program got similar job offers throughout the country." Rebecca interviewed with five companies across the United States. At just twenty-two years old, right out of college, Rebecca was offered a position at a petroleum company that paid $39,000 a year.

Most colleges and universities with accredited programs are visited yearly by recruiters who are eager to hand qualified graduates an appealing employment package. At first, of course, you will have to participate in on-campus interviews during the end of your last semester. Those who do not receive offers from recruiters can still expect to move into their profession relatively quickly. Many smaller companies don't send out recruiters but instead advertise at the university placement center, as well as in trade journals, in professional association newsletters, and on the Internet. A small percentage of entry-level jobs in chemical engineering are found in newspapers.

If your chemical engineering program in college requires you to take an internship, this can be an excellent opportunity leading to full-time employment after you graduate. Rebecca did an internship at Mobil Oil in Atlanta and was offered a job when she graduated, but she had the luxury to decline because it was not in the area she wanted to specialize in. Rebecca does feel that her experience at Mobil Oil was the key to landing her current job.

Chemical engineers are employed by a variety of industries, including the petroleum industry where Rebecca works, and chemical, pharmaceutical, food, electronic, pulp and paper, plastic, rubber, textile, metal, cement, aerospace, and at government agencies, such as the Department of Energy and the Environmental Protection Agency. You can apply directly to specific companies or government agencies by sending them your resume along with a cover letter detailing your skills and qualifications, and why you would like to work for them.

WHERE CAN I GO FROM HERE?

In the private and government sectors, chemical engineers advance based on their qualifications, although to different employers one aspect or area of expertise may weigh more heavily come promotion time. Engineers with only a bachelor's degree execute the experiments, plans, or systems as designed or created by senior chemical engineers. While it is certainly not impossible to advance without more education, it is much more difficult.

Chemical engineers with advanced degrees, on the other hand, are much more likely to direct research projects and quickly advance into managerial and supervisory positions. Generally, the larger the company the greater the degree of specialization required or expected of the engineer (i.e., the more advanced education and hands-on experience) but, as a result, larger companies offer greater opportunities in terms of career growth.

Other chemical engineers teach and/or do research studies at universities and colleges. Depending on the size of the school and the amount of funding devoted to the research project, these positions can be competitive in terms of salaries and benefits.

Rebecca hopes eventually to move into a management position where she will have complete responsibility over individual projects. She is already in a management training program where, under the guidance of her current supervisor, she is learning the skills needed to run larger projects. "I like working with people," she explains. "I know I have the leadership abilities to coordinate a large team. Management is where I'm headed in the next five years."

A small percentage of chemical engineers decide to teach in high schools. While the pay will not be as good, many find it highly rewarding to teach others about the field.

WHAT ARE SOME RELATED JOBS?

The U.S. Department of Labor classifies chemical engineers under the heading Chemical Engineering Occupations. Related jobs include chemical-engineering technicians, chemical-test engineers, absorption-and-adsorption engineers, chemical design engineers, chemical-equipment sales engineers, and technical directors of chemical plants.

Related Jobs

Chemical-design engineers

Chemical engineering technicians

Chemical equipment sales engineers

Chemical test engineers

Technical directors of chemical plants

WHAT ARE THE SALARY RANGES?

The average starting salary for chemical engineers varies depending on what field you are working in. Petroleum tends to pay among the highest salaries, and environmental services tends to pay among the lowest salaries to its young recruits. The general starting salary for chemical engineers in all industries is $39,000. "After my first year I got a 7 percent raise," Rebecca adds. She can

WHAT ARE THE SALARY RANGES?, CONTINUED

expect even higher raises in the next five years as her skills grow and she takes on more responsibilities.

Salaries in engineering often reflect your level of education as well as your years of experience. According to a 1997 salary survey in *Chemical & Engineering News,* those with just their bachelor's degree have a median salary of $59,300. Those with a master's degree earn a median salary of $69,800. The median salary for Ph.D.s is $80,000.

Chemical engineers with a bachelor's degree who have become PEs can climb the salary ladder and over time earn more than $80,000 a year. Salary increases for chemical engineers have consistently stayed ahead of inflation.

WHAT IS THE JOB OUTLOOK?

Because chemical engineers are employed in so many industries, the future is partially dependent on how well each industry performs. In general, chemical engineers can expect a rate of growth as fast as the average for all other occupations through the year 2006. Best bets for chemical engineers in the coming years are bioengineering, microelectronics, specialty chemicals, and plastics. Petroleum and mining have limited futures, as these resources become depleted and new technologies or materials replace them. However, there are fewer students entering petroleum studies, which could create a demand in the near future for qualified people to fill replacement positions.

Traditionally, chemical engineers have worked in manufacturing. Future growth, however, is expected in nonmanufacturing industries. More chemical engineers will be taking jobs in service industries, and will find work as consultants, in environmental services, in law, medicine, and academics. There will still be plenty of positions opening in manufacturing for entry-level workers and those with more experience.

Electrical and Electronics Engineers

SUMMARY

DEFINITION
Electrical and electronics engineers are concerned with developing practical applications of electricity in all its forms. Electrical engineers work with "heavy current" electricity in developing equipment and processes that produce and distribute electricity, such as systems that generate high-power electricity. Electronics engineers work with "light current" electricity in developing virtually anything that uses electricity, from a computer to a camera to a satellite.

SALARY RANGE
$39,513 to $60,100 to $80,000+

EDUCATIONAL REQUIREMENTS
Bachelor's degree

CERTIFICATION OR LICENSING
Voluntary

EMPLOYMENT OUTLOOK
Faster than the average

HIGH SCHOOL SUBJECTS
Computer science
Mathematics
English (writing/literature)

PERSONAL INTERESTS
Building things
Computers
Figuring out how things work
Fixing things

It's Friday afternoon and Peter Kraus listens through his headphones to the prerecorded part of the presentation he's about to give. It's time to introduce the project he's been working on for the past two years to his co-workers, bosses, and a few potential clients. He's tired and tries to stifle a yawn. Last night he was up late running numerous simulations on his computer to make sure the digital signal processing application would work properly. Some circuit operations had been giving him trouble in the past and he wanted to make certain that he had the problem fully solved before he gave his demonstration.

Peter has selected several musical examples to play around with to show everyone how the system he designed will provide superior sound reinforcement capabilities. He turns the volume up from the control console, sets equalization levels on his computer, and rapidly enters various commands into the computer. The presentation plays, and judging from the smiles that wash over the faces of his audience, Peter knows his project is a success.

WHAT DOES AN ELECTRICAL/ELECTRONICS ENGINEER DO?

Think of the last time your neighborhood experienced a power outage. Things you took for granted such as storing perishable food or heating or cooling your house became impossible. You couldn't even read a book or turn on the television to watch your favorite show. It was pretty difficult to get things done, right? Electricity is truly the blood of modern society. And electrical and electronics engineers are the people who keep this electrical "blood" flowing as well as devise new and creative uses for it in our society.

Electrical and electronics engineers use their skills in and understanding of the sciences, engineering, and electricity to develop practical applications of electricity in all its forms. Modern life is unthinkable without the work of electrical and electronics engineers. Their creative ingenuity powers the world. And things aren't changing anytime soon. With the development of new technologies, the widespread use of electronic devices is becoming increasingly prevalent. From pagers to grocery checkout scanners to your neighborhood street lights, this technology is a priceless part of our lives, thanks to the work of electrical and electronics engineers.

The difference between the two types of engineers, electrical and electronics, is basic. They are distinguished by the comparative strength of the electric currents they work with. Electrical engineers work with "heavy current" electricity in developing equipment and processes that produce and distribute electricity, such as systems that generate high-power electricity. Electronics engineers work with "light current" electricity in developing virtually anything that uses electricity, from a computer to a camera to a satellite. However, with recent technological advancements, the line between the two branches is becoming thinner, as each branch uses elements of the other to operate.

Electrical and electronics engineers (EEEs) work in many different industries and perform a variety of tasks. Some of the more

Lingo to Learn

Ampere: *The unit of measurement for an electrical current.*

Capacitor: *An electrical element used to store electrical energy, consisting of two equally charged conducting surfaces having opposite signs and separated by a dielectric.*

Circuit: *A connection of two or more electrical elements to serve a useful function.*

Dielectric: *An insulator, or nonconductor of electricity, usually referring to the insulating material between plates of a capacitor.*

Generator: *An electromechanical device used to convert mechanical energy into electrical energy.*

Radiation: *Energy emitted as electromagnetic waves or subatomic particles from a source or substance.*

Semiconductor: *A substance with electrical conductivity, such as germanium or silicon, which is between being a good conductor and a good insulator used in such devices as transistors, diodes, and integrated circuits.*

Voltage: *The energy required to move a specific number of electrons across two points.*

Watt: *The unit of measurement for electrical power.*

common industries where EEEs work are consumer electronics, electric power, aerospace, computers, communications, and even biomedical technologies. They work in developing things as large as power plants to provide electricity to entire regions of the country, and as small as microprocessors to make your computer run more efficiently. EEEs can be part of a research and development (R&D) team that invents new products and ideas, or part of the design team that actually takes a researcher's concepts and designs the actual product using a computer. Electrical and electronics engineers can also be involved in the actual construction and operation of a design project.

Most engineers choose to specialize in a certain area of electrical and electronics engineering. A complete list of specialties could fill up the rest of this chapter, but some common areas are computer and information engineering, signal processing, electromagnetic compatibility, acoustics, geoscience and remote sensing, lasers and electro-optics, robotics, ultrasonics, ferroelectrics, and biomedical engineering.

Many electronics engineers specialize in computer engineering—the fast growing area of this branch. Computer engineers design the guts of the newest, fastest computers. They use their skills in electronics to develop computer circuitry, information and communications systems, and fundamental computer science applications.

WHAT IS IT LIKE TO BE AN E/E ENGINEER?

As project engineer for a professional sound equipment company, Peter Kraus has many responsibilities throughout the day. While working on multiple projects in various stages of completion he must be sure to keep track of the needs and challenges of each individual project. In doing so, Peter works an average of fifty hours a week. One of his major projects consists of working on ways to improve the company's sound reinforcement equipment. "This is basically a high-tech sound system in a computer," Peter says. "I specialize in digital signal processing, which, in a nutshell, changes analog recording to digital and then back to analog in the computer. This bypasses the complexities inherent in analog circuitry. It's pretty high-tech stuff," he says, with a satisfied smile.

Peter spends his mornings at his desk reading and responding to interoffice correspondences and emails from clients. "I have to set aside a part of my day to do this," Peter says, "because I get so many questions throughout the day that if I answered them when I got them, I'd never get anything done." Peter stresses how important it is that he devote serious time to his co-workers' and clients' inquiries. Sometimes he can reply with a short response and be done before 10:00 AM; other times he has to do a little research or prepare tests to answer their questions, which can last well into the afternoon. "I don't get down to real engineering work until I've handled everybody's questions," Peter explains. Once a week he holds meetings with each team he's managing to get reports on progress, boost morale, if it's needed, and give instructions.

Peter spends a lot of his time working on a computer during the initial phases of a project. He uses the computer to perform complex math equations, do schematic drawings, run simulations of designs, and for many other applications specific to digital signal processing. "Some days I don't leave my desk," he says. "It can get a little frustrating sitting here all day, but with the nature of my work time goes by so fast I hardly notice that it's time to leave. Some days I get so involved in a simulation plan or a really crazy math calculation that by the time I look up from my desk it's already eight o'clock."

Right outside Peter's office is a lab filled with electronic devices. During various stages of a project Peter may have to work with equipment like an oscilloscope, signal generator, soldering iron, and power supply generator. When running simulations on his computer, he periodically must go into the lab and perform tests on the equipment to see how certain circuitries react under specific conditions. "The piece of equipment I'm running tests on this week," Peter explains, "is something only I and my team know how to use. We developed it. It's very specific to our signal processing work."

Before beginning certain projects Peter visits the sites where his equipment will be installed. "Many of the units we design are what we call site-specific. This means we design them with a particular environment with very specific features in mind." Peter's sound reinforcement equipment is used in theaters, music clubs, churches, and by touring entertainment companies. After the system is installed, which generally takes two years from the time Peter first visits the site, he returns with his development team to see how his equipment works in use. "It's great," Peter says. "I've been to all kinds of live concerts and theaters for free. It's very satisfying to be at a club or theater where the sound system sounds awesome and I can say, 'I'm responsible for how good this sounds.'"

Another time that Peter gets out of the office is during the actual production of the equipment. He goes out to the factory to give input, advice, and general instructions to the assembly workers. "I take my work very personally," Peter explains. "It's really easy to make a small error when setting up production. This can screw up everything, so I just like to make sure it's all being designed precisely to my specifications. Most assemblers appreciate my input. I mean, I know better than anyone how it's supposed to be put together."

HAVE I GOT WHAT IT TAKES TO BE AN E/E ENGINEER?

Accuracy and precision are the qualities intrinsic to success as an electrical/electronics engineer. This means attending to details, whether the project is a microscopic computer chip or the largest motorized telescope in the world. "My day is filled with problems to solve," Peter says. "If my team and I don't pay attention to every detail, if we don't check and recheck our work, if our test equipment isn't calibrated perfectly from the onset, we can run into some serious and expensive problems later on down the road." Besides the obvious, professional reasons why precision and accuracy are important in today's fast-paced world, there is an even simpler, more practical reason for crossing every 't' and dotting every 'i'—many projects involve high-voltage electricity; any error in precision or accuracy resulting in a malfunction also has the potential to be life-threatening. It certainly helps if the electrical/electronics engineers are analytical and systematic in their application of knowledge, from the routine and mundane tasks all the way up to the more extraordinary and unique of experiments.

"We love challenges—that's why we're all engineers," Peter says. "My work is constantly testing my skills, keeping me smart, and holding my interest in electronics and acoustics engineering." A natural curiosity and interest in science, electronics, and electricity will, of course, pay off in the long run for EEEs, especially when a test or experiment has become frustrating and a great deal of patience is required. In those times, the engineer's inquisitive nature will help to keep him or her involved in the project long after someone else might have given up.

Depending on the industry in which an engineer works, he or she may be working with the same project for many months, even years; or he or she may be working on different projects every day. This can affect an individual's outlook, so it's important that EEEs be realistic when choosing their area of expertise. For instance, because new projects are always coming into Peter's office, he always has new designs to work on. Whereas this rapid turnover suits his personality and works to keep him interested in each new project, someone else who likes to hone and refine their work until absolutely nothing else can be changed or improved upon might not enjoy a position with so many different projects. "Although all my projects are similar," Peter says, "each one has its unique quality that keeps my job from getting old. I get out of the office enough to keep from getting the desk blues, and by the nature of my work, I get to listen to music all the time. It's a great job!"

For those students who believe that engineering and other scientific jobs are all about numbers and formulas, they're in for a big surprise. Communicating ideas—whether in the laboratory to other scientists and engineers, or in the boardroom to marketing professionals—is an essential part of the EEE's job. Report- and grant-writing, formal and informal presentations, and concise, daily verbal transactions are all fundamental skills and, while they may not always be advertised as such, they are often what distinguishes a mediocre engineer from a successful one. "I wanted to be a writer as well as an engineer," Peter explains. "I took all kinds of English classes in high school and college."

To be a successful electrical or electronics engineer, you should:

Have good computer skills

Be able to communicate your ideas to others

Be attentive to detail

Have patience and tenacity to solve difficult and sometimes complex problems

Be willing to travel

Finally, the nature of the electrical/electronics engineer's work necessitates that he or she travel to construction sites, laboratories, and/or factories, to name but a few destinations, in order to inspect a machine, meet a client, or gauge the amount of work needed to complete a project.

HOW DO I BECOME AN ELECTRICAL/ELECTRONICS ENGINEER?

EDUCATION

High School

High school classes in mathematics, science, and computer courses are the best way to prepare for further study in electrical/electronics engineering. In fact, the more solid the foundation in the following classes, the easier it will be to assimilate more advanced knowledge later on: algebra, geometry, calculus, trigonometry, physics, chemistry, and computer science. This latter subject is absolutely crucial; the importance of developing top-flight computer skills, from programming to the Internet, cannot be stressed enough. "One thing I would recommend to anyone thinking about a career in any engineering field is to get as much experience with computers as possible," Peter advises.

In addition to these "harder" subjects, the "softer" disciplines of English, art, speech, and foreign language can broaden the student's range of knowledge and potential applications of his or her scientific learning. For example, an electronic device might have a decidedly popular and lucrative application in the arts, say an advanced listening system for public libraries and art museums. The EEE who hasn't been exposed to these venues might never think of applying the new device to that area, thus missing the possibility to extend his development to other disciplines. Also, a candidate who possesses good communications skills is much more likely to advance into supervisory and administrative positions where dealing with clients and the public is a crucial part of the job. Peter, for instance, studied some French in high school and finds it very useful when dealing with his French-speaking clients in Canada.

Postsecondary Training

A bachelor's degree in electrical and electronics engineering is generally regarded as the minimum requirement for positions as professional EEEs. Students should enroll in a four- or five-year degree program at an approved college or university; for information on whether or not a particular program is accredited, write to the Accreditation Board for Engineering and Technology. Currently, there are 150 colleges and universities that offer such programs. Admission requirements generally are strict, with schools accepting only those students with excellent academic records and top scores on college entrance examinations like the SAT (Scholastic Aptitude Test). Competition for admission to the best schools is tough, as is competition for national- and school-sponsored scholarships. Peter received his bachelor's degree in electrical engineering from Cornell University.

First-year students take general science and math classes, like chemistry, physics, and calculus, plus general introduction courses to the field, plus one or more electives. Most colleges and universities require that students take an English composition or technical writing class.

Second-year students typically take more advanced science and math classes like differential equations and numerical analysis, and begin to study some field-related classes such as fundamental electronics, circuit analysis, and electromagnetic practice.

Students in their third and fourth years of study begin to specialize, taking classes such as advanced electronics, quantum mechanics, lightwave electronics, digital information circuits, applied radio wave engineering, electrochemistry, engineering ethics, and image engineering. It's a good idea for students to consider early on in their academic careers the area or specialty in which they are most interested, so that they can begin to tailor their education to this niche as soon as possible.

Some chemical engineering programs require students to enter into an internship that is intended to provide them with hands-on work experience in the real work world. These internships provide students with the opportunity to meet new people and make valuable contacts, as well as gain a deeper understanding of how the industry works. Internships can last anywhere from four months to a year. Students usually are not compensated for their work, except in the form of credit hours, per the requirements of the degree program.

CERTIFICATION OR LICENSING

Certification in electrical and electronics engineering is called licensing. Licensed engineers are called PEs (Professional Engineers), just like medical doctors are called M.D.s. While licensing is generally voluntary, it may be mandatory for some jobs today. Consultants in electrical and electronics engineering are required by law to be PEs. Some employers want to be able to sell their company to clients as one containing only professional engineers. Licensing also assures employers that they meet state legal regulations. Peter is not a PE but he does have a Ph.D. in acoustics from Penn State, which right now in his career serves as an excellent indicator of his qualifications.

Requirements vary by state, but generally it takes about four to five years to become a licensed PE. Many students begin the process while still in college by taking the Fundamentals of Engineering (FE) exam. This eight-hour test covers the basics—everything from mathematics, electronics, chemistry, and physics, to more advanced engineering topics.

The next step to becoming a PE is obtaining four years of progressive engineering experience. Some states require that candidates obtain experience under the supervision of a PE. After four years of experience, candidates can take the Principles and Practice of Engineering (PPE) exam specific to electrical engineering (there's one for almost every other branch of engineering, too). Candidates who pass this final examination are then officially referred to as Professional Engineers.

INTERNSHIPS AND VOLUNTEERSHIPS

While finishing his senior year at Cornell, Peter had an internship at Intel. He worked as a technician under electronics engineers and he says he gained practical, "work-world" experiences to which he wouldn't otherwise have been exposed. Colleges and universities that require an internship as part of their program usually will set up an internship with a local industry. The purpose of these internships, which last between one and two semesters, is to provide students with a "test-run," as well as valuable experiences and networking contacts. Students get the opportunity to see whether the field or specific job is really what they want to be working toward. Often, students redefine their career goals based on their firsthand exposure to the position or workplace. As mentioned earlier, these internships are usually non-paying; instead, students receive credit hours for their work.

WHO WILL HIRE ME?

Peter landed his first job as an electrical engineer at a consumer electronics manufacturing company. Industry recruiters came to his campus during his senior year and held interviews with qualified graduates-to-be. Many colleges and universities with accredited programs are visited yearly by recruiters who are eager to hire qualified graduates. At first, of course, you will have to take part in on-campus interviews during the end of your last semester. Peter says he had three interviews with companies across the United States. "Since I wasn't really tied to any part of the country, I had the luxury of taking the best offer I got," Peter explains.

With electrical and electronics engineers being the largest branch of engineering, accounting for roughly 367,000 jobs in 1996, there are good employment opportunities for young, skilled, and industrious engineers. Those who do not receive offers from the recruiters of large corporations still can expect to move into their profession relatively quickly. Many smaller companies don't actively recruit, but need young, qualified electrical and electronics engineers. These positions can be found in trade journals, at your university placement center, in professional association newsletters, and on the Internet. You'll find only a small percentage of the available entry-level jobs in electrical and electronics engineering in newspapers.

Electrical engineers are employed by a variety of industries, such as manufacturing, where Peter works, computer, aerospace, general engineering and consulting firms, communications and utilities firms, academia, and at government agencies, like the Department of Defense and the Department of Energy.

WHERE CAN I GO FROM HERE?

After graduating from Cornell in 1985 and working in the industry for almost seven years, Peter decided that he really wanted to use his electronics engineering skills in the field of acoustical engineering. He went back to school, this time at Penn State, and earned a Ph.D. in acoustics. "It was a career decision I thought long and hard on," Peter remembers. "As an electronics engineer I got to work with some basic acoustic principles and just became fascinated with it. That's when I knew I needed advanced education." Now, at age thirty-four, as a project manager, Peter has reached a point in his career where he can comfortably stop climbing the career ladder and plan his next career move. His decision will be to either follow the management track—which will require further education—or a technical track where he will get increasing autonomy over projects and people until he eventually becomes head managing engineer.

Most entry-level engineers with bachelor's degrees begin their careers by taking positions as junior assistants with minimal responsibility. They generally report to a more experienced engineer, completing tasks and projects as assigned and gradually accruing experience and increased responsibilities. Simultaneously, they begin to ascend the career ladder. Many EEEs find that their ascension to the more coveted positions is that much more rapid if they return to school for advanced degrees after working for four or five years. With five to ten years of experience and/or advanced training or licensing, he or she can hope to advance to the positions of supervising engineer, chief engineer, or plant manager. Those hoping for top-level promotions are also those who are the most business savvy; candidates for these positions should consider honing their business skills either by returning once again to school for courses in business administration, or by attending lectures and seminars on developing management skills.

Other paths open to EEEs include teaching and consulting work. Many EEEs take teaching positions at high schools, colleges, and universities, enjoying the rewards of sharing their passion for science and electronics with young people. Still others work for themselves or private companies as consultants. Called in to offer the benefit of their expertise on a project-by-project basis, consultants may work for one company or several, for anywhere from a day to several months, even years. Often, consulting work leads to offers of permanent work. EEEs certainly aren't limited to their own specialty; on the contrary, they're unique combination of skills qualifies them to work in many other fields, including environmental control, law, medicine, biomedical specialties, as well as other branches of engineering, like chemical, civil, or mechanical.

What Are Some Related Jobs?

The U.S. Department of Labor classifies electrical and electronics engineers under the heading Electrical/Electronics Engineering Occupations. Some related jobs include instrumentation technicians, power-transmission engineers, printed circuit designers, cable engineers, communications equipment technicians, illuminating engineers, electrical-design engineers, electrical technicians, electrical test engineers, electronics-design engineers, electronics technicians, electronics-research engineers, electronic sales and service technicians, and studio operations engineers.

What Are the Salary Ranges?

Related Jobs

Cable engineers

Communications equipment technicians

Electrical-design engineers

Electrical technicians

Electrical test engineers

Electronic sales and service technicians

Electronics-design engineers

Electronics-research engineers

Electronics technicians

Illuminating engineers

Instrumentation technicians

Power-transmission engineers

Printed circuit designers

Studio operations engineers

The average starting salary for electrical and electronics engineers will vary depending on what field you are working in. The average starting salary for electrical and electronics engineers with a bachelor's degree is $39,513. Yearly bonuses can increase salary even more. Those entering the profession with a master's degree can expect salaries ranging from $42,000 to $48,000. And those with a Ph.D. can expect starting salaries in the upper $50,000s and more. According to a 1997 survey by the Institute of Electrical and Electronics Engineers, the average median income for its members was $72,000.

"I've seen my salary and my colleagues' salaries rise pretty steadily as we gain experience," Paul says. "I just recently got an 8 percent raise." Paul says that project engineers—those at the midpoint of their careers as engineers—can make between $60,000 and $85,000 a year depending on their experience and the field they are working in. EEEs in senior management-level positions average $90,000 annually.

Electrical and electronics engineers can usually expect a good benefits package, including paid sick and holiday time, two weeks vacation (and more, the longer they stay with one company), personal time, medical coverage, stock options, 401 K plans, and many other perks, depending on the company and industry.

WHAT IS THE JOB OUTLOOK?

The job outlook for electrical and electronics engineers is good through the year 2006, especially for those who focus on computer careers. One glance at the computer section of any major city's job pages and you'll find page after page of available positions in software, hardware, networks, systems, and many more computer-related jobs, most of which require some background, if not experience, as an electronics engineer. However, due to the volatile nature of the computer industry, and the long hours many computer engineering jobs demand, turnover and layoffs are common. Those who keep up with new technology will have little problem finding employment.

The job market also looks good for EEEs who do not focus on computer careers. According to the Engineering Workforce Commission (EWC) the number of bachelor's degrees awarded in electrical and electronics engineering has dropped since the mid-1980s. The good news for EEE students is the market for these jobs has increased, meaning jobs are plentiful and starting salaries for graduates are up.

As for any job, electrical and electronics engineers' jobs are affected by the ups and downs of the economy; however, since EEEs work in so many different industries, they have the ability to move more freely should one industry be suffering more than others. Of course, they may have to undertake additional training to get up to speed in the new industry.

As the U.S. market becomes a world market, some EEEs will be affected by jobs moving oversees. Job security can no longer be taken for granted. Recent government cutbacks in defense have meant the loss of jobs for many EEEs in the early 1990s, and serves as a warning of what can happen to jobs in a seemingly strong industry when demand decreases. Most EEEs will have at least one significant job change in their careers.

New careers for EEEs are largely computer-related, and especially Internet-related. The consumer electronics industry is always looking for good R&D electrical and electronics engineers to help forge new products. Those electrical and electronics engineers who work hard, keep up with current trends in the industry, and have an open mind toward change can expect a bright future in their respective fields.

Environmental Engineer

SUMMARY

DEFINITION
Environmental engineers *design, build, and maintain systems to reduce or prevent damage to the environment by municipal or industrial wastes.*

ALTERNATIVE JOB TITLES
Public health engineer
Sanitary engineer
Waste management engineer

SALARY RANGE
$30,000 to $45,000 to $75,000

EDUCATIONAL REQUIREMENTS
High school diploma
Bachelor's degree
Master's degree recommended

CERTIFICATION OR LICENSING
Recommended

EMPLOYMENT OUTLOOK
Little change or more slowly than the average

HIGH SCHOOL SUBJECTS
Biology
Chemistry
Computers
Earth science
English (writing/literature)
Mathematics
Physics

PERSONAL INTERESTS
Building things
The Environment
Science

Standing around an open manhole, a team of environmental engineers and maintenance professionals debate their next step. They need to locate an entrance pipe that feeds into the city's sewer system at this point, but the manhole was built without an access ladder. The hole, which leads down to the rushing sewage stream, is narrow—just four feet in diameter—and quite deep. "Someone has to go down," a member of the crew concludes, "and it's got to be somebody small." All eyes shift to Terry Perry, who is smaller, and easily seventy pounds lighter, than anyone else present.

With a rueful grin, Terry assents to the unspoken query. "Yeah, okay, I'll go down. Just don't drop me."

The team quickly assembles a tripod and winch structure above the manhole. Terry, whose only concession to the proximity of raw sewage is a pair of rubber boots, slips into a harness that is attached, by a thick rope, to the winch. Using a large air compressor, the crew blasts air into the manhole to eliminate the potentially oxygen-deficient atmosphere Terry will encounter.

Flashlight in hand, Terry strides to the rim of the manhole. As the crew begins to lower her down the hole, someone jokes, "Don't worry, Terry. If you fall in, we'll just fish you out at the treatment plant."

WHAT DOES AN ENVIRONMENTAL ENGINEER DO?

Environmental engineers are responsible for the systems that are basic to our survival—clean air and water, and treatment of wastes. Environmental engineering is an exceptionally diverse field. Environmental engineers work in many different circumstances and concentrate on many different challenges. Some develop systems to purify water and wastewater. Others design systems to dispose of hazardous waste. Still others are responsible for developing and enforcing environmental regulations. No matter where they work, however, all environmental engineers use scientific principles to design, implement, and maintain systems that protect or restore the environment.

In order to design systems, environmental engineers must combine knowledge from various disciplines. They must understand biology, chemistry, architecture, and economics. An environmental engineer must, for example, know how various chemicals will behave when released into groundwater or soil, how they might affect living organisms, and how long they may take to degrade. The same environmental engineer must know how to design an effective system for removing chemicals or for preventing them from leaking into the environment in the first place. Finally, the engineer must be able to design cost-effective systems, using materials that are both reliable and affordable.

Lingo to Learn

Biodegradation: *The use of bacteria or other living organisms to decompose contaminants.*

Remediation: *To remedy or redress environmental problems.*

Septic: *Anaerobic (without air) decomposition typically accompanied by an unpleasant odor.*

Because environmental engineers must have such a breadth of knowledge, most specialize in one of the many distinct areas of environmental engineering, including air pollution control, hazardous waste management, industrial hygiene, public health engineering, radiation protection, solid waste management, water supply engineering, and wastewater control. Depending on where they work, environmental engineers may concentrate on regulatory compliance, regulatory testing and enforcement, remediation, or research.

Environmental engineers who are responsible for helping companies comply with environmental regulations design systems that enable their clients or employers to dispose of waste and emissions in a responsible manner. Each environmental engineering challenge is unique, however—there is no "cookie-cutter" solution to waste management problems. Environmental engineers must evaluate the type of waste, risks posed to living organisms, surrounding population, soil and water characteristics, and cost of materials and procedures. In some circumstances, an environmental engineer might decide to incinerate waste.

Environmental engineers responsible for developing and enforcing environmental regulations may conduct research to assess the impact of various chemicals or materials on the environment. They also may be responsible for testing the emissions and waste streams created by companies or communities. An environmental engineer might, for instance, take samples of emissions from a company's smokestack. If these samples contain chemicals that are dangerous to the environment or to surrounding communities, the engineer will order the company to eliminate the harmful chemicals from their emissions.

Some environmental engineers design and build systems to pump the groundwater or surface water (lakes, rivers, etc.) to a community's water treatment plant. Others design and build systems that will remove contaminants from the waters. Filtration is a common treatment process. Environmental engineers must constantly monitor the systems and quality of water to ensure that communities receive safe drinking water.

Environmental engineers who focus on remediation efforts must begin by analyzing the type of environmental contamination and track down its source. This can be a painstaking process. If, for example, an environmental engineer finds traces of a commonly used industrial solvent in a community's water supply, he or she must trace the contamination back to the waste's source. The engineer must then identify nearby companies that might use the solvent. Once he or she has narrowed the possibilities, the engineer might have to test the waste streams of several companies to locate the source of the contamination. When the source has been identified, the company must move quickly to redesign its systems to cease the contamination.

In the meantime, the environmental engineer must wrestle with the problem of eliminating the contamination that has already occurred. Environmental engineers today have many methods from which to choose. Before selecting the appropriate method, an environmental engineer must evaluate the chemical contaminant. Some contaminants degrade quickly and can be allowed to degrade naturally. Others break down into chemicals that are more dangerous than the original contaminants. Once the engineer has selected an effective, safe method for removing the contamination, he or she must design the system, oversee its implementation, and monitor its operation.

WHAT IS IT LIKE TO BE AN ENVIRONMENTAL ENGINEER?

"My days vary," muses Terry Perry. "Some days I sit at my desk and design all day. Other days, I visit construction sites where I'm up to my knees in mud. I wear everything from suits to blue jeans and hard hats. This morning I met with several members of the Army to discuss some work we are doing for them, so today was a 'suit day.' "

As an environmental engineering consultant, Terry usually works on several different projects, for several different clients, at a time. She is often responsible for every aspect of a project, from attracting the client and managing the budget, to designing the system that will solve the client's problem and overseeing its construction. Terry estimates that she spends about 20 percent of her time designing systems. "I have to consider every detail," she explains, "right down to the size of the pipes and how many bolts to use to put them together."

Terry spends another 20 percent of her time in meetings with clients and or colleagues. She spends about 15 percent of her time writing reports and devotes approximately 10 percent of her time to marketing the company's services by writing proposals, visiting prospective clients, and making presentations. These days, Terry spends only about 5 percent of her time at project sites. "We usually ask our newer engineers to spend about 95 percent of their time at the sites, so they can gain field experience. I do still visit sites, though, to oversee contractors, make sure they adhere to our plans, and answer questions."

With a small groan, Terry calculates that she spends the remaining 30 percent of her time on the telephone, talking to clients, staff, vendors, and contractors.

While each environmental engineering job is unique, it is possible to make a few generalizations. Most environmental engineers spend time working both indoors and outdoors. The majority of environmental engineers spend up to three-fourths of their time in an office, though a small number do work primarily outdoors. Most engineers also spend significant amounts of time working on computers. Environmental engineers who work for multinational corporations may travel extensively, while those who work for local government agencies may stay in one geographic region. Most environmental engineers work closely with other professionals, including architects, builders, hydrogeologists, laboratory technicians, and chemists.

"I like my job a lot," says Terry Perry. "I like working with a variety of people, such as architects, technical field people, and clients, and I like solving problems. That's really what environmental engineers do—we solve problems."

HAVE I GOT WHAT IT TAKES TO BE AN EE?

As Terry Perry indicates, environmental engineers must be good problem-solvers. They must be able to organize information and identify relationships between a multitude of details. Environmental engineers are the sleuths of the environmental industry.

To be a successful environmental engineer, you should:

Be organized and detail-minded

Be a good problem-solver

Have good oral and written communication skills

Be mechanically inclined

Have integrity and be committed to doing what's right for the environment

Because environmental engineers usually work within teams of professionals, they also must be cooperative and flexible. They must be able to listen to, and evaluate suggestions from a disparate group of professionals. Excellent leadership and communications skills are essential.

"Environmental engineers must be mechanically inclined," says William "Bill" Anderson, who is the executive director of the American Academy of Environmental Engineers as well as a professional engineer. "Environmental engineers must be able to turn abstractions into reality, so they need an aptitude for working with hardware."

Most importantly, environmental engineers must have integrity. Environmental engineers sometimes find themselves in situations that require them to strike an appropriate balance between achieving an employer's objectives and protecting the environment. "Environmental engineers must be able to work with regulatory agencies to get the best deal for their employers or clients," notes Terry, "but they also must be committed to doing what is reasonable for the environment. Environmental engineers have to be committed to doing what's right."

HOW DO I BECOME AN ENVIRONMENTAL ENGINEER?

Many environmental engineers can, and do, enter the profession immediately after obtaining a bachelor's degree. Because the body of knowledge and the technological capabilities within this field are constantly growing, however, many environmental engineers today choose to pursue master's or doctoral degrees. An increasing number of environmental engineers also opt to specialize in one area of environmental engineering.

HOW DO I BECOME A . . . ?, CONTINUED

EDUCATION

High School

High school students who hope to pursue a career in environmental engineering should concentrate on math and science courses. Algebra, calculus, biology, chemistry, and physics all are important classes for the future environmental engineer. Students should also take as many computer courses as possible. "Students need to understand the logic behind computers," says Bill Anderson. "They need to understand how programs work.

"Environmental engineers operate in a field where nothing is black and nothing is white," Bill continues. "When a software program arrives at an answer, environmental engineers need to know how the answer was achieved. They need to know the judgment parameters within which the software operates and, if necessary, they need to be able to change those judgments to match their own."

Since environmental engineers must draw the designs for various systems, students also may find drafting courses helpful, although the computer has largely replaced the drafting table and instruments. Laboratory classes can help students become familiar with mechanical equipment and develop an understanding of how things are put together. Classes such as English and speech can help students develop the communication skills that are so essential to a successful career in this field.

Postsecondary Training

At the college level, students should major in environmental engineering. They should supplement this course of study with chemistry and biology courses and they should continue to study computer programming. Since environmental engineers often must translate highly technical information into language that nontechnical people can understand, students also should take any available technical writing courses.

After completing an undergraduate program, students should at least consider pursuing a master's degree in environmental engineering. Bill Anderson estimates that 50 percent of all environmental engineers today have master's degrees. "The information and technology relevant to this field are undergoing explosive growth," he explains, "and students simply cannot learn enough in the typical 120 undergraduate credit hours."

INTERNSHIPS AND VOLUNTEERSHIPS

Outside of class, students should look for opportunities to learn more about the profession by interviewing environmental engineers or by seeking unpaid intern experiences. Such volunteer positions are available, but students have to show initiative to find them. Bill notes, "Students can help environmental engineers by conducting research and gathering information. This experience can be extremely valuable for students who hope to enter this field."

CERTIFICATION OR LICENSING

Though licensing is not required, it is extremely valuable for environmental engineers. Licensed environmental engineers have considerably more authority to approve and implement system designs.

In order to obtain licensing as professional engineers, graduates must sit for an eight-hour exam immediately after completing their undergraduate degree. Students who pass this exam earn the Engineer-In-Training designation. They must then complete four years of on-the-job experience before taking another eight-hour test. Engineers who have four or more years of experience and have successfully completed both examinations are licensed professional engineers, and may use the initials PE after their names.

The American Academy of Environmental Engineers offers additional certification opportunities. Environmental engineers can take additional examinations to obtain certification in seven areas of specialization: air pollution control; solid waste management; industrial hygiene; radiation protection; water supply/wastewater; hazardous waste management; and general environmental engineer. Environmental engineers also may strive to become diplomate environmental engineers. Diplomate environmental engineers are those who have eight years of engineering experience, four of them in a supervisory capacity, and have passed a written and an oral examination. Environmental engineers who have met these qualifications may use the initials DEE after their names.

WHO WILL HIRE ME?

According to Bill Anderson, it is difficult to generalize about where environmental engineers work. Because the field has so many possible applications, career opportunities are extremely diverse.

The consulting industry is currently the largest employer of environmental engineers. Environmental engineers who work as consultants typically have a broad spectrum of responsibilities. They help large companies comply with environmental regulations and they design systems for waste disposal. They serve municipalities by developing systems to test and treat water to ensure that the drinking supply is safe. They sometimes help government agencies enforce regulations by testing the emissions or waste streams created by various companies or communities. They also plan, implement, and oversee remediation efforts. Environmental engineers who serve as consultants often are responsible for marketing their services, for maintaining budgets, and for managing vendors and support staff in addition to their engineering responsibilities.

Many industrial manufacturers and utility companies hire full-time environmental engineers to help them comply with environmental regulations, to design operations that will create minimal waste, and to plan and implement efficient remediation efforts. Government entities employ environmental engineers to develop and enforce environmental regulations. The armed forces employ environmental engineers to develop systems to dispose of the hazardous and radioactive waste created by munitions plants.

Academic institutions also hire environmental engineers to teach and conduct research. "There are currently one hundred graduate programs in environmental engineering and thirty to forty baccalaureate programs in the country," says Bill, "and they all need professors."

In addition to the more obvious sources of employment, environmental engineers may find opportunities in a number of seemingly unrelated fields. Legal firms, for example, may hire environmental engineers to review the remediation efforts undertaken by one or more parties to a lawsuit. Investment firms may ask environmental engineers to assess the environmental risks involved in various investments. Companies that manufacture and sell environmental technology may hire environmental engineers for advice about the design of, or need for, various pieces of equipment. Municipalities or highway departments may ask environmental engineers to assess the potential impact of a proposed development on the surrounding environment.

Environmental engineers who are highly motivated and creative may find many unusual, interesting opportunities within this field.

WHERE CAN I GO FROM HERE?

Not surprisingly, environmental engineers can advance their careers and increase their earning potential by obtaining additional education, earning licensure, or becoming certified in one or more areas of specialty. Continuing education is extremely important to advancement in this field because the body of relevant knowledge is constantly growing and evolving.

Environmental engineers can also advance by assuming managerial responsibilities. This is true of environmental engineers who work for government agencies, private industry, consulting firms, and academic institutions. Most employers will offer individuals higher compensation for assuming responsibility for a department's performance, budget, and administrative duties. Because environmental engineers who become managers usually must sacrifice some of the time they might ordinarily dedicate to design and field work, individuals who truly love these aspects of the job may find greater satisfaction in non-managerial positions.

WHAT ARE SOME RELATED JOBS?

The U.S. Department of Labor classifies environmental engineers under design and with workers in civil engineering occupations. Also included in these categories are people who design hydraulic devices, irrigation systems, railroad tracks, water utility systems, landing fields, and highways. Some related occupations include architects, drafters, tool designers, photographic engineers, civil engineers, electro-optical engineers, railroad engineers, mechanical design engineers, hydraulic engineers, transportation engineers, sanitary engineers, ceramic engineers, and wastewater management engineers.

Related Jobs

Architects
Ceramic engineers
Civil engineers
Drafters
Electro-optical engineers
Hydraulic engineers
Mechanical design engineers
Photographic engineers
Railroad engineers
Sanitary engineers
Tool designers
Transportation engineers
Wastewater management engineers

WHAT ARE THE SALARY RANGES?

The salaries for environmental engineers depend on the individual's level of education and experience, and on the type and location of employment. An environmental engineer who works for an industrial giant, for example, usually will earn more than one who works for a local governmental agency. Environmental engineers who work in major metropolitan areas typically earn more than those who work in more rural settings.

WHAT ARE THE SALARY RANGES?, CONTINUED

According to "Careers in Environmental Engineering," a document Bill Anderson prepared in 1994, environmental engineers in nonmanagerial positions who have bachelor's degrees can expect to earn starting salaries between $28,000 and $32,000. Those who have master's degrees typically earn starting salaries of $30,000 to $35,000, and those with doctoral degrees earn starting salaries of $35,000 to $40,000. Environmental engineers' salaries tend to increase rather quickly as engineers gain experience. After five years, for example, a licensed engineer can expect to earn $40,000 to $45,000. Those with twenty years or more of experience may earn between $65,000 and $75,000.

WHAT IS THE JOB OUTLOOK?

"For the past forty to fifty years," says Bill, "this field has been somewhat cyclical. Environmental engineering is driven by regulation. When our country emphasizes the enforcement of environmental regulations, environmental engineers have more work. When enforcement is relaxed, there is less work."

//From a long-term perspective, there will always be environmental engineering jobs because there always will be a need for what we do."

The 1980s were a time of increased environmental regulation and enforcement. Superfund legislation forced states to clean up hazardous waste sites and the U.S. Environmental Protection Agency required companies to reduce waste and dispose of it more responsibly. Environmental engineers, consequently, had abundant opportunities. "The field was booming in the 1980s," Terry notes, "but a lot of the major cleanup efforts are now underway or finished, so the environmental engineering job market is kind of tapering off. The future of the market will depend partly on what new legislation is passed."

The Clean Air Act of 1990 did create a brief surge in air pollution jobs. Overall, however, the water supply and water pollution control specialties presently offer the most job opportunities for environmental engineers.

"From a long-term perspective," says Bill, "there will always be environmental engineering jobs because there always will be a need for what we do."

Mechanical Engineer

SUMMARY

DEFINITION
Mechanical engineers *design, test, build, and maintain all kinds of mechanical devices, components, engines, and systems. One of their chief concerns is the production, transmission, and use of power.*

ALTERNATIVE JOB TITLES
None

SALARY RANGE
$38,113 to $65,000 to $100,000+

EDUCATIONAL REQUIREMENTS
Bachelor's degree; an advanced degree will be necessary to land the best jobs

CERTIFICATION OR LICENSING
Recommended

EMPLOYMENT OUTLOOK
About as fast as the average

HIGH SCHOOL SUBJECTS
Computer science
Mathematics
Physics
Shop (Trade/Vo-tech education)

PERSONAL INTERESTS
Building things
Computers
Figuring out how things work

Carlos Gomez tosses empty two-liter plastic bottles into a large steel basin. He has helped design a machine that will sort the empty bottles, nozzle up, single file down a conveyor belt, to a holding slot, where, for a few seconds, the bottle stops to be filled with a liquid. The prototype machine has been having troubles, causing some of the bottles to get crushed and stuck as they queue up at the top of the conveyor belt. Carlos and his team are at a loss for the reason the problem is occurring. According to their many tests and simulations this shouldn't be happening.

Carlos stands on a ladder to look into the basin. As the mechanism positions the bottles, he studies the movements of the bottles, and waits for "the crunch." "Throw some more in," he shouts above the noise to his design partner. "Can you hand me that wrench, and a Phillips too, please." Carlos removes one of the steel panels and sticks his arms and head deep inside the machine. A few minutes later he emerges, black grease on the sleeves of his white shirt. "It's the belt," he says, relieved. "It's just the belt. It's too high and the third positional arm catches it. I'll need some sockets to fix this," he says, and comes down off the ladder. "Well, there goes another shirt."

WHAT DOES A MECHANICAL ENGINEER DO?

Mechanical engineers design and manufacture a variety of items, primarily tools and machines, which generate, transmit, or use power. The items created by mechanical engineers may be used within their own field, but a great deal of mechanical engineering involves designing tools that other specialist engineers or professionals use in their areas of expertise. The mechanical engineer's work falls into the general areas of energy, manufacturing, and engineering-design mechanics.

Beginning with a basic problem—say, how to transport a great number of people through an urban environment quickly and cheaply, how to deter sea mammals from swimming into commercial fishing nets, or how to dissect a single-nerve cell under a microscope—the mechanical engineer researches the existing strategies of solving these problems to determine if there are alternative solutions. In addition to research, he or she conducts experiments and creates models.

Lingo to Learn

Conduction: *The process by which heat travels through a substance.*

Convection: *The process by which heat is transferred by the movement of fluids.*

Gear: *A toothed wheel, cylinder, or other machine element that when turned fits with another toothed element to transmit motion, or to change speed or direction.*

Generator: *A device, consisting of a magnet and coil of wire, which transforms mechanical energy into electrical energy.*

Mechanism: *A part of a machine containing two or more parts arranged so that motion of one part compels the motion of the others.*

Momentum: *The measurement of the motion of a body, determined by multiplying its mass by its velocity.*

Based on the information gleaned from his or her research, the mechanical engineer next creates alternative solutions, complete with new experiments, models, and formulations to test his or her hypotheses. From these tests come the mechanical engineer's designs for a new, partially-submerged railway, an ultrasonic device that transmits high-pitched sounds to warn dolphins and seals, or a laser-based microscopic scalpel. Often, part of the design process also includes coming up with the means of manufacturing or producing the solution. In addition, mechanical engineers design instruments, controls, and power-producing engines such as internal combustion engines, steam and gas turbines, and jet engines.

Even after designing a new device, tool, or machine, the mechanical engineer again submits his or her creations to further testing. Improvements and modifications to the original design are made, and even more testing occurs, both in the laboratory and in the field.

Once the design or theory has been put through many tests and is reasonably, if not completely, error-free—or "bug-proof"—the mechanical engineer then assists in the manufacturing process. He or she selects the proper raw materials, equipment, machines, and systems necessary to make the best product, whether that product is a railway system, an ultrasonic device, or a micro probe. In addition, the mechanical engineer is responsible for supervising the setup, start-up, and safe operation of the product's manufacture.

Finally, the mechanical engineer oversees the day-to-day operation and maintenance of the manufacturing process; supervising technicians and machine operators, assessing the safety and efficiency of the system, and guiding the repairs and regular maintenance of the machines and equipment.

The above aspects of the mechanical engineer's job could easily be performed by one individual; this is often the case in smaller companies or projects. On the other hand, large manufacturers can more readily afford to employ many individuals with specialized knowledge to work on specific tasks or areas; in this way a research/development team of ten mechanical engineers are all working on alternative fuel sources, while another ten mechanical engineers are devoted to the design of a machine that could deliver an alternative fuel, while yet another ten mechanical engineers are concerned with testing the machines developed by the design team . . . and so on and so forth through the various stages of the manufacturing process—from concept to product. Most mechanical engineers are employed in one of these seven areas of concentration (research, development, design, testing, manufacturing, operations and maintenance, and sales). As these areas tend to overlap, and as it is necessary to share information and ideas in order to solve mechanics-based problems, engineers in each area will work closely with one another to create successful solutions.

Mechanical engineers are employed in industries as diverse as automotive, computers, heating and air conditioning, farm equipment, petroleum, metals, and utilities, among others. Depending on the industry in which they work, mechanical engineers provide a wide array of services; in fact, mechanical engineering is said to be the broadest category of engineering, encompassing the widest variety of tasks and functions. Many mechanical engineers end up specializing in one area, though, such as applied mechanics, design engineering, heat transfer, power plant engineering, pressure vessels, and underwater technology.

What Is It Like to Be a Mechanical Engineer?

Carlos Gomez is a project engineer at a mechanical engineering company that makes a variety of industrial machinery. He is responsible for tracking and keeping the deadline schedules for up to three different projects at a time. On average, Carlos works about forty-five hours a week, but during deadlines and problem times, he may work up to sixty hours a week. "If we have a machine that's scheduled to ship to Europe on a certain day, you'll see the lights on all night in this plant until we've got it working perfectly," he says.

Depending on what project he is working on and its stage of completion, Carlos' daily tasks can vary widely. On Mondays he budgets his time for the week. "How closely I follow my schedule," he laughs, "is a whole other issue. Phone calls, unforeseen problems, late meetings, really anything can change my day." Although Carlos may not be able to follow his weekly schedule as closely as he would like, he stresses it is extremely important for him to set down weekly goals. "As long as I know and my team knows what needs to get done for the week, I can assure you, it will be done before we leave Friday afternoon." Carlos says he has occasionally had to go into the office for a few hours on a Saturday to finish minor tasks.

What Is a Machine?

*What comes to mind when you think of a machine? A car engine? A sewing machine? A movie projector? Indeed, these are all machines, but so is a hammer, a screwdriver, a wheel, a crowbar; anything that helps you do work is a machine. There are six so-called "simple machines" upon which mechanical engineering is traditionally based. They are the **lever,** the **pulley,** the **wheel and axle,** the **inclined plane,** the **screw,** and the **gear.** We use these simple machines alone and combined, forming "complex machines" for five general purposes: to transform energy, to transfer energy, to multiply force, to multiply speed, and to change the direction of force.*

Carlos begins work on a new project by first defining the problem and laying out the design concepts. "What is it we want to build? What is it supposed to do? How should it do this task—in other words, what's its specific mechanical function? How much money do we have to build it? What are the design limitations and unique qualities? I ask all these questions and many more technical ones before I begin a design project," Carlos explains. Together with other engineers, managers, and their clients, Carlos comes up with a written outline of how the project will proceed, its time line for completion, a budget, and rough design sketches of the product.

The next stage of a project is making calculations. Carlos uses Excel (a computer program) to make most of his calculations, and he usually has a couple of regular scientific calculators on his desk as well. "We can get into some pretty scary mathematical mazes," he jokes. "Sometimes I spend the entire day making calculations." Depending on the project, these can include horsepower requirements, gear ratios, stress calculations, or flow requirements. Accurate calculations are needed to proceed with the detailing and specifications of the design. Each part Carlos plans to use must have its dimensions and use "detailed" to see where and how it fits into the design as a whole, and to ensure utmost accuracy in the design.

❝In this business you have to stay up to date with the trends, with the materials, and with the people in the field. You are never through learning new things about mechanical engineering. ❞

The next stage, where Carlos spends most of his desk time, is drafting. He uses CAD (computer-aided design) software to design the machine. "The drafting table is a thing of the past," he explains. CAD allows Carlos to rotate a three-dimensional image on his computer screen and view it from any angle. He also uses it to simulate and analyze the effects of different operating conditions. Using the computer has the advantage over the drafting table by allowing quick minor changes to be made to the design and presenting the results immediately.

Carlos also does a lot of work in the shop. Although he has a machinist to actually build his designs, Carlos still fiddles with them, making them just right. He makes adjustments, checks alternate settings, and performs other assembly work. "I've ruined plenty of shirts," he says. "As a mechanical engineer, you have to be mechanically proficient, you have to be able to turn a wrench, and be willing to get your hands full of grease."

A large chunk of Carlos' time is spent out on service work. As the designer of a machine, he knows it best. When a client has a problem, Carlos has to go out and fix it. His clients are all over the world. "I've traveled to Europe several times, to Southeast Asia, Brazil, and Russia, and many states in the United States," Carlos explains. He estimates he's out of the office about a quarter of the year.

Carlos tries to devote a couple hours a week to research. He reads trade magazines to learn about engineering trends, and he attends conferences or seminars on topics he thinks will be valuable to his work. He talks with his vendors about the parts he will be buying and he tries to learn more about new parts applicable to his type of design projects. "In this business you have to stay

WHAT IS IT LIKE TO BE A . . . ?, CONTINUED

up to date with the trends, with the materials, and with the people in the field," Carlos explains. "You are never through learning new things about mechanical engineering. Something that changes the way you do your job is always just around the corner."

HAVE I GOT WHAT IT TAKES TO BE A MECHANICAL ENGINEER?

An innate sense of curiosity is probably the most essential requirement for mechanical engineers. They work with numbers and theories, so having a compelling interest in the results is key to enjoying the work and being successful at it. Without that natural desire to figure out how things work, the job of a mechanical engineer would quickly grow dry and boring. "You have to have a genuine curiosity for mechanics," Carlos stresses. "If you were the type of kid who took things apart—your bike, new toys, a tape player, whatever—you probably have the makings of a mechanical engineer." Having the skills and patience to put things back together correctly is equally important, Carlos says.

Technically, mechanical engineers also should be the kind of individuals who pay scrupulous attention to their projects, from the slightest detail to the organizing principles. Calculations must be flawless, specifications must be accurate, and all drawings must be complete. "A misplaced decimal point on a drawing or a forgotten line can be devastating to a project," Carlos says. Mechanical engineers need to be willing to devote the extra time it takes to thoroughly and exhaustively research any and all possibilities, and then analyze them for their potential.

To be a successful mechanical engineer, you should:

Have an inquisitive mind

Have good communication skills, both written and oral

Be interested in figuring out how things work

Have an aptitude for math and science

Be attentive to details and a good problem solver

Have the ability to make decisions and stick to them

Second only to analytical skills is the ability to be decisive under pressure. In fact, Carlos notes that one of the toughest parts of his job is to make the correct or best decision about a certain aspect of a project when several viable options are available. "Mechanical engineering is all about making decisions. You can make the wrong decision, the right decision, or a less-than-perfect decision." It's the last of these Carlos fears most, since the results of the less-than-perfect decision are not seen until much later in the production process, often when it will be costly to fix them. "You have to make decisions and stick to them," he says. Confidence helps in this regard, and that is only developed through experience and hard work;

knowing that all possible solutions and strategies have been endlessly reworked and revisited in order to ensure that the final solution really is the best solution.

A knack for resourcefulness is also a crucial quality for the successful mechanical engineer. Many projects are tied to strict budgets, so the mechanical engineer who consistently comes up with cost-effective alternatives is certain to find himself or herself in high demand.

Finally, ideas are only as good as they sound. If you can't communicate your earth-shattering discovery, guess what? It will go unnoticed, or worse, someone else who can better communicate its impact will take all the credit for it. Learn to speak clearly and concisely, and people will listen. Carlos wholeheartedly agrees with this, as he often deals directly with clients, vendors, and workers he is supervising, and needs to be able to make himself understood.

How Do I Become a Mechanical Engineer?

EDUCATION

High School
Students interested in pursuing careers in mechanical engineering should get off to a solid start by taking as many science, mathematics, and computer science courses as are offered, especially algebra, geometry, calculus, trigonometry, physics, and computer science. Mechanical engineers do far less hand-drafting than before, but good drafting skills are still highly recommended. In addition, taking a course in computerized drafting is a good idea. English, literature, and speech classes are also important for building strong communication skills. Knowing one or more foreign languages could feasibly triple the number of jobs available to you.

In addition to school subjects, the potential mechanical engineer should pursue whatever hobbies, groups, or projects intrigue him or her. These very same interests may one day turn out to be his or her life's work; at the very least, a hobby or interest can offer the chance to develop related skills and personal qualities, such as perseverance and precision. Carlos' hobbies in high school were repairing lawn mowers and playing around on his computer. He credits both hobbies as being instrumental to his continuing interest in mechanics and troubleshooting, as well as for teaching him a much-needed quality—patience.

HOW DO I BECOME A . . . ?, CONTINUED

Postsecondary Training

The core of the mechanical engineer's work is based in the mechanical sciences. Each mechanical engineer must have a broad understanding of these sciences, and he or she must be able to apply them to practical uses. The typical course of study includes dynamics, which is concerned with the relation between forces in motion; and thermodynamics, which is concerned with the relationships between the forms of heat, energy, and power. The other areas are automatic control, heat transfer, fluid flow, lubrication, and the basic properties of materials. While these topics and subjects are covered in high school, they are covered in much greater depth in college-level courses.

A bachelor's degree in mechanical engineering is the minimum educational level required to work in this field; it is also the first step toward becoming a professional engineer. Many colleges and universities with programs in mechanical engineering follow an approved course of study as set forth by the Accreditation Board for Engineering and Technology. Programs do exist which are not accredited, however. Potential students of mechanical engineering should verify that the programs which they are considering are accredited. The typical program lasts four years, although some may last longer and/or require that students continue with graduate work in the field. Carlos received his bachelor's degree in mechanical engineering in four and a half years from the University of Iowa.

The first two years of most mechanical engineering programs are typically devoted to mathematics, physics, and chemistry. Students take such math courses as calculus, sequences and series, linear algebra, and elementary differential equations; they also take basic engineering courses like mechanics, thermodynamics, fluid mechanics, material sciences, and electrical sciences.

In the junior and senior years students study advanced mechanical engineering, covering areas such as dynamics, design of elements, heat transfer, computer integrated manufacturing, lasers, pulmonary mechanics, stress analysis, and air conditioning and refrigeration. Many programs require that students also take classes in computer programming. In addition, students are generally asked to fulfill core requirements in supporting fields such as the humanities and communications.

CERTIFICATION OR LICENSING

The certification process in mechanical engineering is called licensing. Licensed engineers are called Professional Engineers (PEs). While licensing is generally voluntary, it may be mandatory for some jobs today. For example, consultants in mechanical engineering are required by law to be PEs. Licensing also assures employers that they meet state legal regulations. Although Carlos is not a PE, he says he's been thinking about it more and more lately. "If I want to move up to chief engineer, I'll have to get my PE," he says. Obviously, licensing is directly tied to higher salaries, better jobs, more responsibilities, and timely promotions.

Requirements vary from state to state but it generally takes about four to five years to become a licensed PE. Many mechanical engineers begin the process while still in college by taking the Fundamentals of Engineering (FE) exam, an eight-hour test that covers everything from electronics, chemistry, mathematics, and physics to more advanced engineering issues.

Once a candidate has successfully passed the FE exam, the next requirement to fulfill is to acquire four years of progressive engineering experience. Some states require that you obtain your experience under the supervision of a PE. Once a candidate has four years of on-the-job experience, he or she then takes another exam—the Principles and Practice of Engineering (PPE)— which is specific to mechanical engineering (each branch of engineering has its own specialized, upper-level test). Candidates who successfully complete this examination are officially referred to as Professional Engineers. Without this designation, mechanical engineers aren't allowed to refer to themselves as PEs, or function in the same legal capacity as PEs.

INTERNSHIPS AND VOLUNTEERSHIPS

An internship may be required as part of your college curriculum. This hands-on work experience at a company employing mechanical engineers will give you a chance to work side-by-side with mechanical engineers on real-life problems and projects. This experience allows students to gain valuable exposure to the field and practical applications of their studies, as well as providing them with future networking contacts. Internships routinely last four to twelve months and are usually arranged for by the school, especially when the internship is a requirement for the degree. They are usually nonpaying, but do count toward semester credit hours.

"I got a good feel for what it was really like to work in the field," notes Carlos of his summer internship with the John Deere Company. During the summer before his senior year, he worked as a technician under the supervision of mechanical engineers.

WHO WILL HIRE ME?

Carlos really covered his bases the second semester of his senior year to ensure that he got a good job when he finished his degree. He signed up for on-campus interviews with recruiters, submitted his resume to the university job placement center, interviewed with a headhunter, wrote cover letters and sent his resume to employers he saw advertising job openings in trade magazines and in newspapers, and networked with the people he met during his internship. "I did more work to get a job than I probably needed to," Carlos says, "but I wanted to know before I graduated that I had a job. I wanted to get in the real world right away." In fact, Carlos began his first day on the job at a small naval design firm only a week after graduation.

In the end, Carlos took the job a headhunter located for him. He also received offers from two of the on-campus recruiters, and he had three interviews with companies that advertised job openings in various trade magazines—one of which made him an offer. He said it was a hard decision but he took the job that seemed to offer the most growth potential.

All large industries employ mechanical engineers, so depending on your interests and particular skills, you have many options. Some of the major industries employing mechanical engineers include automobile, heavy and light machinery, machine tool, construction equipment, transportation, heating and refrigeration, and power and energy, as well as engineering or consulting firms, and government laboratories and agencies.

Jobs for mechanical engineers are usually advertised in the multitude of industry trade magazines. Some major cities have a large engineering jobs section in their classified pages where employers advertise openings. Increasingly, jobs are posted on the Internet. The American Society of Mechanical Engineers offers a large jobs section on their Web site (http://www.asme.org).

WHERE CAN I GO FROM HERE?

Many career avenues are open to the mechanical engineer, perhaps more so than any other engineering discipline. Career paths in mechanical engineering resemble those in other branches of engineering in that mechanical engineers can pursue a technical career in research and development; a teaching and research post at a major research university; a corporate position in the sales department of a manufacturer; or a supervisory/managerial position directing the work of other mechanical engineers in the design and manufacture of products. Clearly, an ambitious candidate in each of these scenarios will rise to these roles with the right education level, experience, and hard, disciplined work, but other qualities can make the difference between a stellar career and a mediocre one; again, it depends on the industry in which the person works *and* the person's area of specialization. For instance, the person with excellent business skills, even an advanced degree in business administration, will be more likely to move into an upper-level position.

Mechanical engineers entering the job market with only a bachelor's degree will find that without further education and training their possibilities for advancement will most likely fizzle out. On the other hand, mechanical engineers with advanced degrees and the professional engineer designation (PE) will qualify for higher-level, supervisory positions.

Carlos hopes that, with five to seven more years' experience under his belt, he will be able to move up to the chief mechanical engineer position. "The chief mechanical engineer," Carlos explains, "is in charge of all the project engineers. He has his hands in all the projects and makes the final decision on many important issues. The job holds a lot of responsibility."

From there, the next step at a large company would be chief engineer. This person looks over all the other chief engineers in their respective fields, for example, mechanical, electrical, plastics, and safety engineering.

WHAT ARE SOME RELATED JOBS?

The U.S. Department of Labor classifies mechanical engineers under the heading Mechanical Engineering Occupations. Workers in this group include automotive engineers, tool and die designers, mechanical engineering technicians, mechanical drafters, plant engineers, numerical control tool programmers, optomechanical engineers, and mechanical research engineers.

As mentioned earlier, mechanical engineers work with other types of engineers in a variety of fields. Some of these workers include aerospace engineers, safety engineers, industrial engineers, welding engineers, structural engineers, transportation engineers, civil engineers, petroleum engineers, and electrical and electronics engineers.

Related Jobs

Aerospace engineers

Automotive engineers

Civil engineers

Electrical and electronics engineers

Industrial engineers

Mechanical engineering technicians

Optomechanical engineers

Petroleum engineers

Plant engineers

Safety engineers

Structural engineers

Tool and die designers

Transportation engineers

WHAT ARE THE SALARY RANGES?

Mechanical engineers are among the highest paid engineers. According to a survey by the National Association of Colleges and Employers, the average starting salary offer for mechanical engineers is $38,113. Engineers entering the field with a master's degree or Ph.D. can earn considerably more. Also, earning your certification as a PE will boost your salary. Mid-level engineers earn between $50,000 and $75,000. Engineers in management positions with twenty-five years of experience can earn $100,000 and more.

Mechanical engineers can usually expect a good benefits package, including paid sick, holiday time, vacation, and personal time; medical coverage; stock options; 401 K plans; and other perks, depending on the company and industry.

WHAT IS THE JOB OUTLOOK?

The future looks good for mechanical engineers; so good that *U.S. News & World Report* recently chose mechanical engineer as a "runner-up hot track job" for its 1998 Career Guide. We live in a service economy, where the population has become more and more dependent upon automated systems. Through the year 2006 employment for mechanical engineers is expected to grow at least as fast as the average for all other jobs. In our fast-paced, highly technical society, all industries strive and compete to be on the cutting edge. They require mechanical and other engineers to help them achieve their high-tech goals.

Recruitment out of college has been exceptionally good for the past several years, reports the Engineering Workforce Commission in an April 1996 report. Most offers to these new mechanical engineers come from the automotive, mechanical consulting, aerospace, petroleum, metals, and chemical industries. Other industries that will rely on the skills of mechanical engineers to keep them on the forefront of technology are transportation, environmental control, bioengineering, energy conservation, and robotics.

The number of bachelor's degrees awarded in mechanical engineering has declined since the mid-1980s, while the market for these jobs has grown. This is particularly encouraging news for new grads in the field. Also, due to the aging population, and older mechanical engineers leaving the work force, more positions are opening than can be readily filled. Most mechanical engineers will have at least one significant job change in their careers.

One setback for mechanical engineers is in the defense industry, where spending cuts have resulted in significant layoffs.

Packaging Engineer

SUMMARY

DEFINITION
Packaging engineers *design the packaging for consumer goods such as food, electronics, medical products, toys, appliances, clothing, and many more. Packages are designed to protect products, provide benefit to consumers, conserve natural resources, and minimize waste through recycling.*

ALTERNATIVE JOB TITLES
None

SALARY RANGE
$34,000 to $66,800 to $94,448

EDUCATIONAL REQUIREMENTS
Bachelor's degree

CERTIFICATION OR LICENSING
Voluntary

EMPLOYMENT OUTLOOK
Faster than the average

HIGH SCHOOL SUBJECTS
Computer science
Mathematics
Physics

PERSONAL INTERESTS
Building things
Computers
Figuring out how things work

Strewn across a folding table are several long narrow boxes, all the same size but different in appearance. Roger Klein and his packaging team have designed them for a client who needs to ship promotional documents to potential customers all over the world. Some of the boxes open with the pull of a cord, others will require a sharp object to open, and one opens with a tear along its edge. Each box is slightly different in color and made of a different material. Two of the team members prefer a plastic one for its durability but Roger prefers a sturdy heavy-grade paper one because it's cheaper and still attractive. "Our client's logo also prints better on the paper," he says, picking up both and holding them side by side.

Roger begins stuffing the two envelopes with samples of his client's material. Immediately he sees a flaw in the paper envelope that will prevent them from using it. "It's going to rip during delivery," he says, frowning. "The sharp corners of this plastic calendar will shift and I'm sure it can put a rip right through." Still, Roger prefers the looks of this envelope. He holds it in his hand and thinks. "What if we double-line the edges with our 8-grade stock?" he asks his team members. They all nod. Someone says, "That'll do the trick." And Roger rushes off to his computer to make the adjustments to his design.

WHAT DOES A PACKAGING ENGINEER DO?

Just about everything you buy in a store comes to you in packaging. It may be a candy bar packaged in some kind of wrapper, a roll of film that's first packaged in a plastic container and then in a cardboard box, or your new stereo system, which usually comes wrapped in some kind of plastic, then tightly packaged between Styrofoam and finally boxed for shipping—whatever the method, most packaging is so common in our daily lives that we hardly pay any attention to it. In fact, the packaging industry has played a significant role in the evolution of living standards by making available a wide variety of products, from perishable foods to fragile electronics, to consumers all over the world.

Packaging engineers take many factors into consideration when designing a package. Their first question is, "What are we going to package?" Due to the vast variety of products requiring packaging, some engineers may specialize in the packaging of liquids, frozen foods, electronics, medicines, hazardous materials, or specialty products. They research new materials and methods for specific types of packaging and build prototypes to test for durability, attractiveness, environmental impact, and many other factors established by their clients and the end user.

In packaging, engineers use four main groups of materials: glass, paper and board, metals, and plastics. They study ways these materials can be used together, and how each material can be altered for new applications. Increasingly the packaging industry has become sensitive to the environmental repercussions of its packaging, and has taken steps to ensure the safety of the environment.

An engineer's chief concerns when designing packaging are the needs of his clients and the consumer's expectations. For example, a client may need a complex container for shipping a delicate piece of equipment, or a consumer may expect a package of cookies to open and then seal in a particular way. Generally, clients will specify their packaging needs to be either for shipping, storage, display, protection, or to fulfill some unique customer expectation of a product, such as child

Lingo to Learn

Packaging: *All products made of any materials to be used for the containment, protection, handling, delivery, and presentation of goods from the producer to the user or the consumer.*

Primary packaging (also known as sales packaging): *The packaging of a product to be a sales unit at the point of purchase. For example, a bag of potato chips.*

Secondary packaging (also known as grouped packaging): *The packaging of a single product in a group in such a way that removal of the packaging does not affect the product's characteristics. For example, a case of cola.*

Tertiary packaging (also known as transport packaging): *The packaging of a product to facilitate handling and transport of multiple sales units. It is also designed to protect the product and primary and secondary packaging during transport. For example, a cardboard box case of 100 tubes of toothpaste.*

safety guards built into medicine bottles. Consumer expectations are usually determined through extensive marketing surveys.

In designing a new package, engineers use computer software such as CAD (computer-aided design) to draft numerous three-dimensional possibilities. When using new materials, they make sample packages to test for the quality and functionality of the material. They must test a variety of materials and designs to come up with a final plan that fits a client's budget, as well as their shipping, storage, display, and protection needs. Sometimes the best plan will not be accepted by a client because it does not fulfill an important requirement. For example, the packaging of a certain frozen food product may fit the client's budget, it may display and ship nicely, but if it does not protect the food against "freezer burn" it is ultimately useless to the client.

Packaging engineers are not only concerned about the package itself, but also with three operations basic to the packaging system, namely producing, filling, and closing the package—all steps crucial to the success of the package and the product. Some packaging engineers, trained in mechanical engineering, are responsible for designing the machines that either make the package or package the product. Most large-scale packaging productions will be handled by a specially designed machine or machines. Engineers take the finished package design and design a packaging process that will fill and close the package. For example, a cookie manufacturer must have a system in place that sorts the cookies, drops or places them in a container, which then must fit in a bag or box, which in turn, must be sealed. And due to the delicate nature of cookies, this whole process must be done without damaging them. Next, the manufacturer must have a system to package large quantities of the smaller packages of cookies so that they can be distributed to suppliers and sent on their way to the consumer—and the whole time the packaging engineer must take into consideration countless factors that can damage the product.

WHAT IS IT LIKE TO BE A PACKAGING ENGINEER?

As a packaging design engineer for a folding carton design firm, Roger Klein has many duties that keep him busy throughout the day. He designs everything from specialty mail envelopes to containers and display boxes for frozen foods to large "packaging systems" designed for large products with many parts, such as a children's home playground set. "For that playground set I had twenty different boxes and retaining chambers built into one large box," Roger remembers. "That set took a lot of time to package." Roger is responsible for tracking and keeping the deadline schedules for up to three different projects at a time. On average, Roger works a standard forty-hour week, but during deadlines and problem times, he may work up to fifty hours a week. "During the design of the playground set packaging I logged several hours overtime to get it done in time for the Christmas season," Roger says.

Depending on what project he is working on, and in what stage of completion it is in, Roger's daily tasks vary. He says there is no such thing as a typical day. "That's one of the things I love about my job," he explains. "My days are always different. There are always unique challenges, and different tasks that keeps a day from becoming routine." With a design taking anywhere from a few hours to a few days to finish, the turnaround time is fast.

❚❚If it's my design, I kind of like to be around while they're coming off the belt. I like to check for quality, appearance, and to just be there if any problems arise."

However, there are typical tasks Roger performs during the week. He spends time each morning going through his email and other correspondence he received late the previous day or overnight. Most of the email pertains to current projects, which Roger must answer immediately. He also receives numerous questions from clients about already completed projects. "The playground set client emailed me recently about some changes they've made to their design," Roger explains. "He wanted to know if the existing packaging would work, because they still have a lot of it left. After doing some calculations on my computer I had to email him back that it wouldn't work."

Inevitably, Roger spends a good part of his day in the CAD department working on new designs. "We have separate computers beefed up to run CAD that we keep in a different area," he says. Using this specialized software Roger creates and alters three-dimensional versions of his packaging designs according to specifications either established by his clients or by earlier tests with prototypes. Roger also frequently designs the dies that will be needed to make the packaging. "I simply send our die maker my design on computer file and in a couple weeks we get our die and install it in our machines. Then production on the package can begin."

Depending on the type of packaging, some are made in-house, and other, usually larger designs, are made at an off-site location. When the actual production of a package begins, Roger spends much of his day in the warehouse where the machines make the packages. "There are usually a few problems that need troubleshooting the first day of production," he explains. Roger works with machinists to coordinate production schedules and plans. "If it's my design, I kind of like to be around while they're coming off the belt. I like to check for quality, appearance, and to just be there if any problems arise."

Some of his day he spends in his office where he has a different computer to do calculations, work on his spreadsheets, write business correspondence, and call his customers and vendors. "Unlike most jobs," Roger explains, "I come to my desk to relax."

About once a week Roger visits new clients or goes to his vendors to find out about new products. He sets up meetings with his vendors to establish costs on certain materials. When he finishes a package he brings it directly to the client. "Most of our clients are in the Midwest, not more than a day trip from the office, but sometimes I have to travel out to the East Coast to present a package," Roger says.

HAVE I GOT WHAT IT TAKES TO BE A PACKAGING ENGINEER?

One of the fundamental responsibilities of a packaging engineer is to scout out problems before they occur, and they do occur. Packaging engineers are faced with countless research, development, educational, strategic, and technological problems when designing a packaging system. For instance, the packaging engineer assigned to create food packets for interspace consumption by NASA astronauts would want to brainstorm any and all possibilities for how a single packet might accidentally tear and leak the contents into the zero-gravity atmosphere in order to devise protective measures against such an occurrence (and still ensure that an astronaut could easily open the packet at mealtimes). If the packaging engineer didn't envision all possible scenarios, from best-case to worst-case, the next moon mission might very well be compromised because the packets broke and the astronauts ran out of food.

Not all packaging situations involve such life-and-death consequences, but if a milk carton leaks or the jewel case around your favorite CD breaks, as a consumer you're not pleased; and you might very well choose another brand the next time. These are just some of the considerations that packaging engineers face. To do their best, they need to pay close attention to the details of all aspects of their projects, including the client's demands, the consumer's habits, the method of handling, and the transportation of their packaged product. No detail is too small to take into account.

Successful packaging engineers may develop expertise in one area, but they continually look for ways to apply their knowledge beyond the boundaries of their specialty. Versatility and flexibility are necessary qualities to achieving this. These specialists design packaging that will protect the product, itself, as well as consumers of the product. Packaging engineers constantly update their knowledge of materials and process in order to design creative solutions that will meet the many different specifications of clients who, at least theoretically, could come from any area or industry. For example, a packaging engineer might be asked to design the lightweight mailer in which a computer software company will ship program disks to its software designers overnight. In this case, both the weight and cost of the package is extremely crucial. On the other hand, the same packaging engineer might be asked to design a protection sys-

To be a successful packaging engineer, you should:

Have a knack for solving problems

Be attentive to detail

Be able to meet deadlines and work in stressful situations

Keep abreast of current technology and industry innovations

Have strong oral and written communication skills

tem for packing and transporting priceless artworks. In this case, weight and cost are no longer primary design concerns; instead, the objective becomes safeguarding the contents from light, breakage, and fluctuations in temperature. "There is absolutely no room for mistakes," Roger says. "If the microwaveable carton we designed for a certain food product is too small or too big, tapered wrong, too thick, too thin, really in any way other than indicated

on our specs, and we've already produced a couple hundred thousand of them, then we're in deep water. That would probably be the last job we did for that client."

Packaging engineers are responsible for designing materials which are also safe for the environment. This means they should have a demonstrated interest in the environment and what their products do to the environment as they are produced and once they have been used and discarded. To understand developing materials and technologies related to the production of eco-friendly materials and processes, packaging engineers should zealously pursue leads on new technologies and materials by reading and keeping up to date on industry trends and forecasts.

Also, packaging engineers must have flexible, adaptable personalities and work well under pressure and in high-stress situations. Often, a concept which worked well one day will be tossed out the next because of new considerations. Packaging engineers must be able to "go with the flow" and entertain new ideas, theories, and formulations as they are created, incorporating them into their overall design as necessary. "Our industry is about change," Roger says. "Many of the materials and processes we use today were unheard of ten years ago. You have to adapt well to change, because the industry is always looking to better itself." Roger notes such significant industry changeovers as the shift from steel to aluminum cans, glass to plastic production.

Every project brings with it a host of problems. The challenge comes in identifying and resolving problems in a timely fashion. This means having the analytical skills necessary to identify problems, and the problem-solving skills to distinguish between small glitches and major catastrophes; situations that can be resolved in the long-term, or crises that need immediate attention before the project is ruined. Prioritizing the concerns of both the client and the engineering teams at work on a project can alternate between science and art.

Packaging engineers frequently work with clients and designers to create a shape or texture that reflects the product's primary appeal to the consumer. To accomplish this, packaging engineers need to be skilled communicators, adept at summarizing why particular concepts won't work and others will.

Finally, packaging engineers need to be able to come up with safe, efficient designs which are cost-effective for both the client and the consumer. Cost-cutting skills are thus extremely important, as is being materials-savvy—knowing which cheaper materials or processes can be used to achieve identical or similar effects. Of course it's possible to create a leak-proof container that keeps milk cool without refrigeration, but can it be done for less than a penny per unit? "Management and engineers frequently have meetings to look at ways of cutting costs without losing quality. It's a real challenge," says Roger.

HOW DO I BECOME A PACKAGING ENGINEER?

EDUCATION

High School
Math and science courses are, of course, integral to any engineer, as are courses in mechanical drawing and computer science, but other, "softer" courses contribute to the educational background of packaging engineers. Courses like

art, computer graphics, public speaking, and English are valuable to the engineering student because they allow the student to polish his or her visual and language skills, both of which are crucial to expressing oneself. For example, packaging engineers are closely involved with many other professionals in the design and manufacture of a product's packaging, including the all-important client. Being able to use the language and metaphors of both scientific and non-scientific professionals not only eases the communication process, but demonstrates a broader, more sophisticated knowledge and perspective. In addition, those with "people-friendly" communication skills are much more likely to be hired or promoted into supervisory and administrative positions.

Prospective packaging engineers also should not hesitate to develop their interests in other areas, as packaging engineers work in almost every industry. One student's interest in cooking might obviously lead to work in an industry related to food technology, whereas another student's seemingly unrelated interest in mythology might lead to an innovative, highly original package design. Other options for exploring areas related to engineering and packaging include joining a science club or an entrepreneurial/business organization like Junior Achievement (JA). The latter organization, for example, offers budding packaging engineers the opportunity to explore business concerns firsthand, as well as develop a product from concept to reality, including delivery to the consumer.

Postsecondary Training

Entry-level positions as packaging engineers require a bachelor's degree. The most valuable education comes from a program which has been approved by the Accreditation Board for Engineering and Technology; students considering applying to programs in packaging engineering should verify with the schools that they are, in fact, accredited programs. Engineering programs of all kinds typically last four years, but others may require additional years of study or a fifth year of practical experience in an internship. Roger's bachelor's degree in packaging engineering from Michigan State University, for example, took four years to complete.

At this time, only a handful of schools offer a degree in packaging engineering. Students at colleges that don't offer a degree in packaging shouldn't despair; many of the same courses are taken by engineering students in all fields, and it is possible to supplement a degree in, say, mechanical engineering, with additional coursework in packaging from another school that will qualify the student for entry-level packaging jobs. If packaging degrees are not offered, the best course is to study a related engineering field and then follow

Packaging Origins

Surprisingly, the packaging of commercial products did not begin until the late nineteenth century. In 1894 a group of U.S. bakers formed a national organization to market a new kind of cracker. They had grown tired of hearing complaints from customers about how fast the thin crackers would go stale, and they wanted to be certain their new cracker was enjoyed fresh. They had also figured fresh, clean, and crisp crackers would result in better sales. To solve the problem, they came up with a folding carton with a waxed inside lining. They sealed the carton with a printed wrap that identified the product and baker. Sales were outstanding, and soon they began designing packages for other products. The labels on the packages established trademark names and increased name recognition. Almost immediately, other producers of goods from syrup to silk stockings began to package their products.

up with an advanced degree in packaging. This route also ensures that the candidate will be considered for better-than-entry-level jobs, as well as higher salaries and benefits because of his or her advanced degree.

The first and second years of the four-year bachelor's degree are spent studying mathematics, physics, and chemistry. Students also take such courses as plastics technology, fundamentals of packaging, and elements of food processing, in addition to basic engineering courses like mechanics, thermodynamics, fluid mechanics, material sciences, and electrical sciences.

Juniors and seniors study courses in advanced packaging engineering, such as packaging materials, food engineering, statistical quality control, packaging in society and the environment, protective packaging, and packaging machinery. Most programs require students to take classes in computer programming. If the program requires an internship, it is usually scheduled during these last two years, as well.

CERTIFICATION OR LICENSING

As with most people whose work involves the public's health and safety, packaging engineers are encouraged to become registered or licensed. In many ways, this process of registering or licensing practicing engineers is a good idea. Companies are required to meet state and national guidelines, which include the number of workers with and without the proper credentials who they are allowed to employ. Lucrative government contracts, for example, often go to those companies who strictly observe licensing protocol by employing only licensed or registered engineers. On the other hand, the packaging engineer with the proper licensing is more likely to be hired at a higher level than those packaging engineers without the proper licensing or registration. In fact, unlicensed packaging engineers are also often ineligible for the better-paying, supervisory or administrative positions. Therefore, in the world of engineering as in other areas of industry, licensing acts as an incentive for both employer and employee.

HOW DO I BECOME A . . . ?, CONTINUED

In engineering, the licensing process results in the formal designation of Professional Engineer (PE). Specifically, packaging engineers must be graduates from an approved engineering school, have four years of experience, and pass a state examination. It isn't necessary that packaging engineers be licensed as such; they may be licensed as materials engineers, or as engineers in another specialty. The important thing is that, as licensed professionals, they have passed standardized examinations which demonstrate their knowledge of the regulations governing their field.

Requirements vary from state to state but generally it takes about four to five years to become a licensed PE. Many engineers begin the process while still in college by taking the Fundamentals of Engineering (FE) exam, an eight-hour test that covers everything from electronics, chemistry, mathematics, and physics to the more advanced engineering issues.

Once a candidate has successfully passed the FE exam, the next requirement to fulfill is to acquire four years of progressive engineering experience. Some states require that packaging engineers obtain experience under the supervision of a PE. Once a candidate has four years of on-the-job experience, he or she then takes another exam specific to packaging engineering (almost every branch of engineering has its own specialized, upper-level test). Candidate who successfully complete this examination are officially referred to as Professional Engineers. Without this designation, engineers aren't allowed to refer to themselves as PEs, or function in the same legal capacity as PEs.

INTERNSHIPS AND VOLUNTEERSHIPS

Packaging engineering programs that require an internship usually either have an established internship program set up in partnership with local companies, or the program's placement department works with the student to locate a suitable internship that will fulfill the requirements. Required internships generally last between one and two semesters and offer no pay; instead, the experience counts for a stipulated number of credit hours, in accordance with the degree program's requirements.

One of the reasons why many degree programs are so involved in the internship aspect of the program is that they want to guarantee that each of their students is in a positive learning environment, taking full advantage of the practical, on-the-job experience to augment their classroom studies. Students should always check with their program administrators if they want to obtain credit for their intern experiences, as some programs are extremely finicky about extending credit for internships. Usually, the company offering the internship must prove to the school that the student is actually learning valuable skills and gaining on-the-job experience, instead of just working for the company for free. Internships solicited by the student (and which are not required by the degree program) may, in fact, offer some form of stipend in addition to, or in the place of, counting toward credit hours at a university. Students can work in these internships during the summer.

WHO WILL HIRE ME?

For Roger, a career in packaging was an easy decision. In high school he worked at a carton plant with die makers and carton designers and got early exposure to CAD software. His employer saw potential in him and encouraged Roger to pursue a career in packaging. In college Roger had a natural head start among his classmates because of his previous work and excelled in his classes. During his senior year he had employers practically knocking down his door with job offers. "Recruiters came to our campus," Roger remembers, "and I had six different offers. My old employer for whom I worked on breaks and the summers also offered me a job." Roger said that there is always a demand in the industry for good CAD operators with creative design skills—his specialties—and that any graduate with these skills will have little trouble finding a job in the industry if it continues its relatively rapid growth.

As the nation's third largest industry, the packaging industry plays a significant role in our economy. The world budget for packaging material and machinery has reached over $300 billion. Unique to packaging engineers is that they can find work not only in the packaging industry, working for such companies as Ball Corporation, Aluminum Company of America, American National Can Company, and Tenneco Packaging, but also in most of the world's largest industries such as food, automotive, pharmaceutical, chemical, cosmetics, electronics, and agriculture. There are also numerous smaller packaging companies that provide specialty packaging for an array of diverse products.

Jobs for packaging engineers are usually advertised in the multitude of industry trade magazines and packaging magazines such as *Packaging Digest*. Some major cities have a large engineering jobs section in their classified pages where employers advertise openings. Increasingly jobs are posted on the Internet. The Institute of Packaging Professionals has a jobs section posted on their Web site (http://www.pakinfo-world.org).

WHERE CAN I GO FROM HERE?

Almost without exception, all products need or benefit from some type of packaging—whether that packaging functions to advertise, protect, and/or transport the products. As a result, packaging engineers work in nearly every industry, in almost every country in the world. "My first job after graduation was in the food industry," Roger says. "We packaged fruits and vegetables. Then I advanced to my current job in the packaging manufacturing industry. I started as an upper-level CAD operator and over the course of six years moved up to senior design engineer." Roger says that further advancement would probably mean shifting focus to management-type positions and would require more education on his part outside of work. "Basically, with my skills and experience I can get a job as a design engineer at any company. My employer recognizes this and pays me well."

Entry-level engineering positions in packaging usually consist of more generic tasks completed under the supervision of one or more experienced packaging engineers; that is, the new, less-experienced packaging engineers execute the research, plans, testing, and/or production of packages in accordance with the directions of the senior, supervisory engineers. Many industries sponsor formal education as well as informal, on-the-job training programs in which experienced engineers basically train the younger engineers to do routine work. As the entry-level engineers gain more experience and demonstrate that they have mastered new skills, they are given more difficult work and greater responsibilities. As they begin to assume an increasingly independent role, they might be responsible for developing designs or even be placed in charge of smaller accounts.

Built into many industries is a hierarchy of jobs. New hires advance sequentially through any number of positions according to their job performance, work experience, educational background, and professional demeanor. For example, a recent packaging graduate might start as an assistant to product designers by creating engineering layouts, advance to product designer, and then move into a management position. Factors affecting the rate of advancement frequently include licensing, advanced degrees, and demonstrated successes in the field, say, receiving an award for package design. Many packaging engineers who decide early on that they want to be involved at the management level go on to take additional courses or obtain advanced degrees in business administration or a related, specialized field. In addition to the formal education engineers can obtain from colleges and universities, many also participate in less formalized educational settings by attending lectures, seminars, and conferences on packaging design, environmentally friendly production processes, et cetera.

In addition to working in the world of industry, packaging engineers also teach in high schools, colleges, and universities. Many are content sacrificing the higher pay they might make in industry for the challenges and rewards of teaching others about the field itself, or one of its primary subjects, such as chemistry or physics.

Finally, packaging engineers aren't limited to their own branch of engineering, but can also find employment in the mechanical, chemical, or materials engineering fields. In fact, they aren't even limited to the field of engineering; many packaging engineers find satisfying work in advertising, law, and computer programming. The educational foundation provided by most engineering degrees allows its students to apply their knowledge in a myriad of interesting ways.

WHAT ARE SOME RELATED JOBS?

The U.S. Department of Labor classifies packaging engineers under the heading Occupations in Architecture, Engineering, and Surveying, Not Elsewhere Classified. Some related jobs include test technicians, materials engineers, biomedical equipment technicians, optical engineers, ordnance engineers, pollution-control engineers, laser technicians, laboratory technicians, logistics engineers, project engineers, design drafters, and specification writers.

Related Jobs

Biomedical equipment technicians

Design drafters

Laboratory technicians

Laser technicians

Logistics engineers

Materials engineers

Optical engineers

Ordnance engineers

Pollution-control engineers

Project engineers

Specification writers

Test technicians

WHAT ARE THE SALARY RANGES?

The starting salaries for entry-level junior engineers in packaging range between $34,000 and $43,000, depending on the level of education (bachelor's or master's degree) and type of industry. A 1996 survey by the Institute of Packaging Professionals indicated an overall mean salary of $66,800 for packaging engineers. The same survey shows a $94,448 mean salary for engineers with corporate management functions.

A recent survey conducted by *Packaging World* found that the highest-paying salaries in packaging came from the pharmaceutical industry. The food

WHAT ARE THE SALARY RANGES?, CONTINUED

and beverage industry also pays its packaging engineers a competitive salary. Senior engineers earn on average $46,500 a year, and chief engineers with over fifteen years experience can earn $67,000 a year and more.

Packaging engineers can usually expect a good benefits package, including paid sick and holiday time, two weeks vacation, personal time, medical coverage, stock options, 401 K plans, and other perks, depending on the company and industry.

WHAT IS THE JOB OUTLOOK?

The future is bright for packaging engineers. Being the third-largest industry, packaging practically has an engine that runs itself. Technological advances in the last five years have sent the packaging industry on an upward climb. Efficiency, quality, and worker productivity have all improved. Many other industries rely on packaging for their products to sell. Packaging engineers know better than anyone that a product's packaging is often one of its selling points—a nifty-looking bottle of mineral water or a handy resealable bag of chips are often the sort of things that set one product's sales ahead of another's. By continually testing new products, researching new materials, and designing different and better ways to package the world's products so that ultimately they sell better, the packaging industry is paving the way for a healthy and productive future.

Currently there is a larger demand for entry-level packaging engineering jobs than there are young recruits. For the recent packaging engineering graduate this means higher starting salaries. "At our firm we just hired two new junior CAD technicians with design skills, both with packaging degrees, and they're making what I was making only five years ago, and I've been with this company for twelve years," Roger says. "That's how fast salaries have climbed."

A current trend in the packaging industry is industry consolidation. Many smaller packaging companies are being bought out by the larger ones. The smaller companies are restructured, which results in an initial job loss, but once operations are worked out and production strategies get underway, jobs are added. Large companies, however, tend to use more automation in their production, which can also displace some workers.

Plastics Engineer

SUMMARY

DEFINITION
Plastics engineers *create, design, and test polymeric materials to manufacture useful end products, from plastic automobile parts to biodegradable polymers for a packaging company to plastic fibers for clothing.*

ALTERNATIVE JOB TITLES
Chemical engineer
Polymer engineer

SALARY RANGE
$35,000 to $64,000 to $100,000+

EDUCATIONAL REQUIREMENTS
Bachelor's degree

CERTIFICATION OR LICENSING
Voluntary

EMPLOYMENT OUTLOOK
Faster than the average

HIGH SCHOOL SUBJECTS
Computer science
Mathematics
Physics

PERSONAL INTERESTS
Building things
Computers
Figuring out how things work

Shelly Davis is nervous. Her heart races as she reads over her presentation for the tenth time in two hours. The current speaker is almost finished with his speech, a seminar organizer tells her, and she will be next. Shelly looks into the audience. There are over a hundred people sitting at tables, taking notes, and paying close attention to the presentations. Everyone, Shelly notices, seems much older than she. Her nerves flare up even more and she takes a deep breath. This is her first presentation. She is confident about her research, her boss has praised her for her hard work, and she knows her facts are accurate and well presented. She hears her name called, and reluctantly she steps up to the podium.

Shelly does not look into the audience but only at her paper. She pretends she is reading it to her research team back in the lab, and slowly her mind eases with each sentence. She has prepared some large charts to illustrate aspects of her research and as she explains one of the charts she sees people nodding with interested agreement. The presentation lasts half an hour and she concludes to the sounds of applause. Afterwards, Shelly is approached by a number of her industry elders who congratulate her on her work.

WHAT DOES A PLASTICS ENGINEER DO?

Taking apart and rearranging the molecular structure of a wide variety of raw materials such as coal, wood, natural gas, or salt, in order to create completely new synthetic materials is the primary work of plastics engineers. They work in many areas from research to design to production, using their scientific knowledge of plastics, or polyethylene terephthalate (PET), to formulate new products or new applications of plastics technology.

Plastics are used everywhere—in the clothes we wear, the cars we drive, the buildings in which we work and live, to name but a few of the more common applications. "The future is in plastics," Dustin Hoffman's character was told in *The Graduate*, and that future is here and now. Either as a part or ingredient, or as the entire product itself, the manufacture of plastics has escalated since the late nineteenth century when John W. Hyatt first created celluloid as his entry in a contest to invent substitutes for the ivory used in billiard balls. Today, synthetic polymers—chains of hydrocarbon molecules—represent a multi-million dollar business as either the main ingredient or the item itself in building and construction, clothing, packaging, aerospace, and consumer products. In addition, plastics has had a stunning effect on the automotive, biomedical, communications, electrical and electronic fields, in some cases breathing new life into them. Since plastic is man-made, relatively inexpensive to produce, highly durable, recyclable, and does not drain our natural resources, one of its most common uses is in replacing products and parts made of traditional materials like wood, metals, and glass.

First of all, the role of a plastics engineer differs according to the industry in which the person works. Plastics engineers work in all phases of polymer production and application, such as raw material supply, machinery manufacture, compounds and converters, fabrication, and the manufacture of products with plastics components. Then, like most engineering jobs, the plastics engineer's job can be broken down into the further categories of research, design, development, manufacturing, quality, marketing/sales, and information.

Lingo to Learn

Additive: *A compound or substance added to a polymer at some point during its processing to affect a desired change in the polymer.*

Grades: *Refers to polymers that belong to the same chemical family, and are produced by the same manufacturer.*

Monomer: *Molecules of an organic substance that are the basic structural unit of polymers. Monomers must be bonded together to form polymers.*

Plastic: *A synthetic or naturally occurring organic substance, which at some point in its formation or manufacturing process becomes formable or pliable. Plastics are generally made up of polymers.*

Polymer: *A substance formed by a chemical reaction in which chemical units (mers), join together in a line to form repeating smaller units. Polymers joined together (with other substances) are what make the varieties of plastics.*

The day-to-day duties or tasks which a plastics engineer completes are based upon the area in which he or she works (although a plastics engineer's responsibilities may cross into with other areas). For example, a plastics engineer working in the *research* division for a chemical company might be striving to perfect a design for a polymer that safely captures and decomposes toxic chemicals (like those used in chemical warfare) without harming the human body or the environment. Such an innovative application of a polymer could potentially put an end to one of the most damaging effects of war—the pain and sickness from chemical exposure experienced by field soldiers since World War I, when mustard gas was first used. Research specialists work with the basic, chemical structure of molecules in order to create new materials.

Once the polymer has been perfected, its creators might work with other plastics engineers in the *design* division to craft delivery methods for the new polymer; that is, they would brainstorm different ways of introducing the polymer to the user, such as in the form of a lotion, an inhalant, or a mask or some other type of protective clothing. Once one or more of these has been produced, they then test them in order to determine the most effective method of introducing the polymer. *Application engineers* develop these new processes and materials to improve existing or newly created products.

After the best delivery system has been designed and tested, the *development* team of plastics engineers get to work figuring out how to create the polymer quickly and easily using the most up-to-date technology. *Production engineers,* or *process engineers,* then take that process one step further as they work to move the production process from the laboratory to the manufacturing plant where, instead of say, one batch of polymer, they will attempt to produce ten thousand batches. It is their job to make the necessary modifications to the formula or process to allow for the mass production of high quality, reliable, predictable materials or products.

Quality control is a vital component in this process, as *quality engineers* are not only responsible for ensuring that the products are consistently of the highest quality, but responsible for seeing that the production process is safe and efficient for those working in the plant, as well as for the environment. The product—no matter how harmless it ends up being to the environment in its final state—is of no value if, during the production process, other toxins or dangerous elements are created or introduced to the environment in the form of waste. Likewise, the product can't be considered a success if it can't be recycled, or easily broken down once its purpose has been served.

WHAT DOES A PLASTICS ENGINEER DO?, CONTINUED

The plastics engineer's role isn't finished even when the product has been safely mass-produced. On the contrary, because of the highly technical nature of plastics products, the skills and knowledge of the plastics engineer are extremely useful in selling and marketing the plastics product or application. Those who go into *sales and marketing* jobs have excellent written and verbal communications skills and enjoy working with clients, suppliers, and manufacturers; and illuminating and explaining the process to other professionals, regardless of their scientific or non-scientific backgrounds.

Finally, plastics engineers offer their unique knowledge and perspectives to others through *teaching* or *consulting* positions. Many plastics engineers enjoy academic careers which range from two-year technical schools to advanced degree programs at colleges and major research universities. On the flip side, many industries with plastics-oriented products frequently employ plastics engineers to act as consultants on those products or projects which involve polymers. They may also suggest new products or design changes to better fit customer preferences and needs.

WHAT IS IT LIKE TO BE A PLASTICS ENGINEER?

Shelly Davis is a research engineer for a large chemical company, whose main business is producing plastics for parts product manufacturers. She is responsible for coming up with new kinds of plastics and, in some cases, new applications for existing plastics her company has designed. She works on one or two projects at a time, some of which can take up to a year of research. "I work with some pretty complicated stuff," Shelly explains, "and there are so many things that can go wrong that sometimes it just takes a while to get it right." In a typical week, Shelly works between forty-five and fifty hours.

Depending on what stage of the project she's working on, Shelly spends some of her day in the laboratory where she performs chemical experiments and tests, or at her desk where she works on her computer. In the laboratory she wears a long, white lab coat, goggles, a hair net, and special gloves. "Some of the chemicals we use are dangerous to the skin and eyes," Shelly explains. "And the hair net and lab coat are to ensure that none of our hair or clothes fibers get into the chemicals. Purity is a big concern."

In the laboratory Shelly mixes chemicals either by hand or using special computerized equipment. She observes and takes notes on chemical reactions. Because some of the results she is interested in are not observable to the naked eye, Shelly views these reactions under the microscope. She also studies

Know Your Plastics!

Polyacrylamide: *A tough clear plastic that compact discs are made from.*

Polyethylene: *The soft clear kind of plastic that typically plastic baggies are made out of.*

Polystyrene: *A stiffer, often white plastic, that your milk is "bottled" in.*

the results of certain processes via microscopes. "I can spend hours bent over the microscope observing and taking notes," she says. "If I don't see the result I'm looking for, I simply make adjustments to the equation and perform the test again. It can get a little boring." Shelly says there is also a physical stress on the eyes and shoulders from looking through the microscope for long periods of time.

Other tests involve observing how certain monomers, polymers, and plastics react to lights, heat, sound, various fluids, and physical pressure. Depending on the research she is conducting, Shelly performs countless tests using a variety of specialized machines. "I subject polymers to radiation, to extreme heat, to extreme cold—you name it. We have a machine that can create a condition to test our plastics and polymers against, to see how they respond," Shelly says.

In addition to laboratory research Shelly does a lot of book research. Because the plastics and polymer industry is growing so rapidly, there have been many technological advancements over the years. Most successful research is eventually published for other engineers to learn from. "Some days I can spend a whole morning or afternoon reading up on what's new in the industry," Shelly says. "It is absolutely crucial to my job as a research engineer to keep as up to date as possible." Shelly also attends industry seminars where new plastics/polymer information is presented for the first time.

Once Shelly and her team have completed a research project, they must create a detailed report of their findings. These are often very long documents of a highly technical nature complete with complex chemical diagrams and mathematical equations, various graphs, charts, and illustrations. The typical report takes Shelly and her team between four and six weeks to assemble. Some of the reports may be submitted to technical journals or even presented at an industry seminar.

HAVE I GOT WHAT IT TAKES TO BE A PLASTICS ENGINEER?

With new polymers being developed daily, plastics engineers are constantly under pressure to integrate new technology and science. Having the imagination to consider all of the possibilities and then being versatile enough to adapt one application of a polymer to another situation are, perhaps, the most essential qualities for plastics engineers. "We get work done by questioning things we take for granted, and by pushing ourselves to go beyond what we think we understand," Shelly says. To accomplish this, plastics engineers must first learn how the new polymer may be applicable to their industry or product line, and then decide how to adjust their current manufacturing process to incorporate it.

To be a successful plastics engineer, you should:

Understand the properties of plastic materials

Be able to adjust and adapt well to changes in the industry

Have an inquisitive mind and an eye for details

Have strong oral and written communications skills

Be able to work as part of a team

In addition to having a good mechanical aptitude for developing plastic parts and tooling, one of the more basic qualities for any student considering a career in plastics engineering is a solid understanding of the properties of plastic materials; their characteristics before, during, and after the process; and their performance in the final application or product.

As in every scientific endeavor, there are always a varying number of factors which influence the outcome of the experiment, and the chemical configurations in polymer construction are no different. It takes an individual with an extraordinary amount of patience, focus, and determination to notice precisely what factors are achieving the desired results. Successful plastics engineers pay attention to the smallest detail, note the nuances between experiments, and then use that information to develop further tests or theories. Having a certain amount of critical distance helps plastics engineers step back from the minutia and reassess the direction in which they're headed. "There are days I get to work and I really don't know what to do first," Shelly explains. "I have to take a deep breath, look over my notes from the day before, and try to take a step back to look at the big picture."

❞We get work done by questioning things we take for granted, and by pushing ourselves to go beyond what we think we understand.❞

Plastics engineers need to be inquisitive, to take creative steps toward improvements by constantly asking questions, and taking a fresh look at familiar practices.

Good communication skills are vital for success in engineering. Shelly, for example, must write reports about her projects from the perspective of the various stages involved in her experiments. The first report indicates her objectives and plans for the project; subsequent progress reports detail any strategy changes or problems (the number of progress reports depends on the duration of the project); and a final comprehensive report summarizes the final results, indicating whether or not the project met its original goals. In addition, she sometimes presents her results before a large audience at industry seminars.

How Do I Become a Plastics Engineer?

EDUCATION

High School

While few courses at the high school level are directly related to plastics engineering, the basic foundation for engineering includes a wide range of math and science courses. Students interested in pursuing a career in this field should invest in an education steeped heavily in math and science, including geometry, algebra, trigonometry, calculus, chemistry, biology, physics, and computer programming. Plastics engineers who will also be designing products will need drafting skills, so mechanical drawing and art classes are an excellent choice. English, speech, and foreign language classes develop strong communication skills and provide students with the opportunity to learn how to better express themselves in both speech and writing.

Ancillary interests should not be overlooked. In addition to providing students with possible ways of applying their scientific knowledge in enjoyable, recreational activities, exploring personal hobbies can also develop crucial personal and professional qualities and skills, such as patience, perseverance, and creative problem-solving. One of Shelly's hobbies in high school was chemistry. She played an active role in many school and community science fairs, in one of which she won a prize for a fertilizer she developed that grew her softball-size tomatoes, she says, insisting that it was her blossoming interest in chemistry early on that helped her develop troubleshooting skills and patience with experiments.

Postsecondary Training

The minimum education requirement to get a job as a plastics engineer varies according to the industry and the plastic processes involved. For many positions, it is possible to have only a high school diploma, while other positions require advanced degrees. One reason for the discrepancy in education backgrounds is that, compared with other branches of engineering, there are relatively few plastics engineering degree programs. Many plastics engineers are actually trained as chemical or mechanical engineers and later take training seminars in plastics specialties. In fact, it has been estimated that of the total number of college graduates currently working in the plastics/polymer industries, only a small percentage, perhaps no more than 10 percent, had had formal training in either plastics or the polymer sciences. Since most companies don't require a four-year degree for project engineers, experience becomes the key factor in landing the better jobs.

Although they may be few in number, excellent degree programs in plastics engineering do exist. The typical course of study for the first year includes basic math and science courses such as physics, chemistry, calculus, engineering, as well as an introductory course in materials. In the second year, students study polymer materials, plastics processes, organic chemistry, economics, safety, thermodynamics, and differential equations. In the third and fourth years, students delve deeper into the focus of their majors, studying fluid flow, process control, fundamental electricity, heat transfer, safety, plastics molding, production process design, polymer structures, and advanced safety, among others.

Plastics: Earth-friendly?

Indeed, plastics are an earth-friendly material. It takes less energy to convert raw material to plastic than other usable products. Plastic saves fuel consumption because it is lighter in weight than metals and glass; and because plastic is durable it nearly eliminates breakage during shipping, a common occurrence for glass-shipped products. And most plastics are recyclable!

Most plastics engineering programs require that students participate in an internship program with a company employing plastics or chemical engineers; some programs may require you to take two internships. This will give you a chance to make valuable contacts as well as learn more about the field from experienced plastics or chemical engineers. Internships can last anywhere from four months to a year.

CERTIFICATION OR LICENSING

Certification and licensing for plastics engineers is not universally encouraged or enforced, unlike other branches of engineering. In many ways, this process of registering or licensing practicing engineers is a good idea, but in plastics engineering, it usually depends on the company and its level of involvement in plastics processing. Undoubtedly, as more and more schools offer degree programs in plastics engineering, there will be a greater need for certification; it will help the industry recognize the benefits of hiring those who have acquired both experience and education.

In general, the licensing process for all branches of engineering results in the formal designation of Professional Engineer (PE). It isn't necessary that plastics engineers be licensed as such; they may be licensed as materials engineers, or as engineers in another specialty. The important thing is that, as licensed professionals, they have passed standardized examinations which demonstrate their knowledge of the regulations governing their field.

Requirements vary from state to state but generally it takes about four to five years to become a licensed PE. Many engineers begin the process while still in college by taking the Fundamentals of Engineering (FE) exam, an eight-hour test that covers everything from electronics, chemistry, mathematics, and physics to the more advanced engineering issues.

Once a candidate has successfully passed the FE exam, the next requirement to fulfill is to acquire four years of progressive engineering experience. Some states require that plastics engineers obtain experience under the supervision of a PE. Once a candidate has four years of on-the-job experience, he or she then takes another exam specific to plastics engineering (each branch of engineering has its own specialized, upper-level test). Candidates who successfully complete this examination are officially referred to as Professional Engineers. Without this designation, engineers aren't allowed to refer to themselves as PEs, or function in the same legal capacity as PEs.

INTERNSHIPS AND VOLUNTEERSHIPS

Schools with degree programs that require an internship usually will have a partnership set up with local industry. Internships, although generally non-paying, do count for semester credit hours. The typical internship lasts between one and two semesters.

WHO WILL HIRE ME?

During her senior year at the University of Massachusetts, Lowell, Shelly signed up for interviews with on-campus recruiters from local and state-wide industry. "I had six interviews," Shelly remembers. "I could have had more, but I chose the top six because interviewing can be pretty stressful."

Shelly also responded to some job postings on the Internet and in trade journals. In all, she had ten job interviews and three job offers. The job she took was for a small plastics company outside of Boston. "From the time I began interviewing to when I got the job, seven months passed," Shelly says. "It was extremely stressful, because I was trying to finish my studies and get a job at the same time. For the job I did get I had three interviews." Shelly finished her degree while still in the interviewing process, and moved back to her parents' house while she waited to begin her new job.

Upon graduation most plastics engineers go to work for industry. Some may continue their studies and go on to teach in higher education. Most plastics programs have advanced programs for master's and doctoral studies. In industry, plastics engineers fall into five main employment groups: manufacturing (where the products are made and tested), material applications and development (where Shelly is), machinery/equipment (which requires advanced knowledge of mechanical engineering), government positions, and consulting (where you will need your Professional Engineer licensing).

Plastics engineers can find work in a variety of industries outside of plastics. There are currently around eight thousand U.S. companies—employing roughly 1.2 million workers—that rely on plastics in whole or part for their business. Typical industries employing plastics engineers are transportation, automobile, construction, packaging, housewares, furniture, optical goods, and electronics. Trade magazines for these industries are good places to look for career opportunities. Because of this diversity of industries, plastics engineers have the luxury of finding employment in all fifty states; the top five being California, Ohio, Michigan, Illinois, and Texas. South Dakota, Utah, and Colorado have recently enjoyed high growth in plastics employment.

WHERE CAN I GO FROM HERE?

Perhaps the best aspect of plastics engineering is that those who work in this field can easily migrate into other branches of the field of plastics without too much difficulty, as the different areas share the same skills and knowledge.

In general, advancing through the ranks of plastics engineers is similar to other disciplines. Working in entry-level positions usually means executing the research, plans, or theories which someone else has originated. With additional experience and education, plastics engineers begin to tackle projects solo or, at least, accept responsibility for organizing and managing them for a supervisor. Those plastics engineers with advanced degrees (or, at this point in time, a great deal of experience) can move into supervisory or administrative positions within any one of the major categories, such as research, development, or design. Eventually, those plastics engineers who have distinguished themselves by consistently producing successful projects, and who have polished their business and managerial skills, will advance to become the directors of engineering for an entire plant or research division.

Currently, Shelly is a research engineer. As part of a four-member team, she predicts that with three to five more years experience she can advance to project engineer. As a project engineer she would be the top engineer in charge of a research team. Her responsibilities would include overseeing all research and plan a work schedule with precise objectives. When problems arise, the project engineer makes the decision of what to do next. "That is the kind of responsibility I want," Shelly says. "I know how to troubleshoot, so that when we run into problems I can solve them. But now I must first talk with my project engineer. He always agrees with my decision." Project engineers typically advance by overseeing multiple projects as project manager.

WHAT ARE SOME RELATED JOBS?

The U.S. Department of Labor has not yet classified the career of plastics engineer. Some related careers that involve tasks, duties, and processes that are somewhat similar to that of plastics engineers include automotive engineers, mechanical engineers, industrial engineers, electrical engineers, materials engineers, biomedical engineers, chemical engineers, environmental engineers, civil engineers, and safety engineers.

Related Jobs

Automotive engineers

Biomedical engineers

Chemical engineers

Civil engineers

Electrical engineers

Environmental engineers

Industrial engineers

Materials engineers

Mechanical engineers

Safety engineers

WHAT ARE THE SALARY RANGES?

Engineers of chemical and allied products (which includes plastics) are among the highest paid engineers in the engineering professions. The salary of entry-level positions will vary depending on the industry and can range from $31,000 to $45,000. The high-paying industries for plastics engineers are chemical, petroleum, transportation equipment, and food/beverage/tobacco. Some of the lower-paying industries include paper and paper products, automotive parts, and fabricated metal products.

Engineers in management positions with nine to eleven years' experience can earn between $60,000 and $68,000. Those plastics engineers with twenty-five or more years of experience can earn $100,000 or more annually.

WHAT ARE THE SALARY RANGES?, CONTINUED

An advanced degree, whether that is PE licensure, a master's or a Ph.D., translates into a higher salary. Those engineers at larger firms average higher salaries than those at smaller firms, with the median income for engineers with one to four years' experience at larger firms at $45,000, and at smaller firms, $39,222.

Plastics engineers can expect a good benefits package, including paid sick, holiday, vacation, and personal time; medical coverage; stock options; 401K plans; and other perks, depending on the company and industry.

WHAT IS THE JOB OUTLOOK?

The outlook for the plastics industry is very good. With 20,000 U.S. facilities producing plastics materials, products, and equipment, amounting to a $4 billion annual trade surplus, the industry contributes significantly to the nation's economy. According to the Society of the Plastics Industry, Inc., plastics outpaces most other manufacturing industries in the number of jobs it creates. From 1991 to 1994 the industry experienced 16 percent growth (and this during a time when our nation was in a recession). The Society of Plastics Engineers says growth in the industry has been unprecedented, and they expect increased growth well into the next century. The U.S. Bureau of Labor predicts that by the year 2000, plastics production will be three and a half times the volume of all metals.

Our natural resources have reached dangerously low levels, and plastics and polymers are increasingly taking their place in many of the products we use. Due to recent technological advancements in the industry, the biomedical, communications, and electrical industries have been turning to polymers to solve some of their problems.

"There is a big buzz in the industry right now," Shelly says. "Jobs are opening up, research is up, our bonuses are up. Things look bright. Right now we have five job openings we're trying to fill." The plastics industry is on a campaign to get more of its entry-level positions filled by engineers who have completed a plastics/polymer curriculum. Until recently, not many engineering programs offered a plastics/polymer bachelor's degree program, making qualified entry-level engineers hard to find. To counter this problem, the industry has started to support existing programs with financial and equipment assistance, and they have encouraged other engineering schools to create plastics/polymer programs.

Transportation Engineer

SUMMARY

DEFINITION
Transportation engineers *plan, design, and operate all methods, structures, and systems that transport people and goods in a safe, convenient, rapid, and environmentally responsible manner. Typical projects they work on are streets, highways, tollways, airports, transit systems, railroads, and harbors.*

ALTERNATIVE JOB TITLES
Civil engineer
Traffic engineer
Transportation planner
Transportation professional

SALARY RANGE
$37,740 to $66,000 to $93,000

EDUCATIONAL REQUIREMENTS
Bachelor's degree

CERTIFICATION OR LICENSING
Voluntary

EMPLOYMENT OUTLOOK
About as fast as the average

HIGH SCHOOL SUBJECTS
Computer science
English (writing/literature)
Geography
Mathematics

PERSONAL INTERESTS
Building things
Computers
The Environment
Figuring out how things work

Phil Hartley leans over an aerial photograph of a congested urban highway area where three different highways merge, which has been a high accident area for years. Phil is busy trying to find a solution to the problem. Using special geographic computer software, Phil zooms in on an off-ramp where many rush hour backups and accidents occur. His fingers move rapidly over the keyboard as he enters the keystrokes. His eyes flash continually from the photograph to the computer screen. Phil thinks he sees a solution and then the phone rings.

A city highway transportation official wants to know if they can meet tomorrow to discuss possible solutions. He tells Phil the city is expecting some large events in the coming years and they hope to have the problem solved before then. Without taking his eyes off the computer screen, Phil schedules their meeting for the following afternoon. He jots the time and place down quickly on his calendar and returns to his monitor. He has several calculations to make and computer simulations to run on his idea before he makes his recommendations; still, he feels confident he sees a solution that will help reduce traffic as well as fit in the city's budget.

WHAT DOES A TRANSPORTATION ENGINEER DO?

Transportation engineers play a very important role in society. Wherever you go, whatever roads, bridges, and walkways you take to get to your intended destination are a result of the work of a transportation engineer. They plan, design, and operate all means of transport, including streets and highways, transit systems, airports, railroads, shipping ports, harbors, and bicycle paths. In designing these methods of transportation, they are constantly considering such aspects as safety, efficiency, comfort, convenience, economy, and the environmental friendliness of each project. The six major areas that transportation engineers work in are: highways, airways, railways, waterways, conveyors, and pipelines.

Lingo to Learn

Capacity: The maximum number of vehicles expected to pass a specific roadway in one direction in a specific time period under normal roadway and traffic conditions.

Channelization: The dispersement of conflicting traffic movements into separate paths of travel by traffic islands or pavement markings to ensure the safe and orderly movement of vehicles and pedestrians.

Dwell time: The time, in seconds, that a transit vehicle is stopped to serve its passengers.

Infrastructure: The life-support facilities of geographic areas such as roadways, water facilities, sewers, harbors, and pipelines.

Sight distance: The length of roadway ahead visible to the driver.

Terminals: The points where travel and shipment begins or ends.

Traffic flow: The movement of people and goods in an urban area.

Volume of flow: Number of vehicles observed to pass a point during a specific time.

Zoning: Provides the assurance that land uses of a geographic area are compatible in relation to one another by controlling and setting restrictions on the type of use in the area.

Transportation engineering is a very diverse field. It is a multidisciplinary area of study within engineering as a whole and closely linked with civil engineering. Because of this inherent diversity, transportation engineers use applications from many other fields in performing their varied tasks. In addition to other branches of engineering, such as civil, electrical, mechanical, and industrial transportation, engineers draw on concepts from geography, economics, regional planning, statistics, sociology, and many other fields when doing their work.

Some transportation engineers work in planning. When a project begins, transportation planners meet with other transportation professionals, engineers, neighborhood groups, and government officials in order to learn of and address the concerns and challenges a project may have. Typical concerns and challenges include environmental, wildlife, landscaping, special design needs, appearance, and the community impact of a project. Planners for a rapid transit system determine the course of the tracks, where the stops will be, and ease of access by the general public. Planners often must defend or justify a proposed project before neighborhood groups,

Fast Facts

business leaders, and the media. They often must write long, detailed reports explaining a project or defending it. Planners work closely with government officials and environmental groups to ensure all needs and concerns are addressed.

Design engineers design the methods of transportation for unique situations. For example, they may be challenged to build a highway off-ramp that must avoid a wetlands preserve or enter a suburban area with minimal noise disturbance. Designers build rapid transit systems in already crowded cities, and must pay close attention to overall cost and efficiency of their design. Most transportation engineers use computer programs such as CAD (computer-aided design) when designing new transportation facilities.

Operations is a very important field in transportation engineering; with safety being one of its largest concerns. These engineers are responsible for how traffic flows, whether the traffic consists of automobiles, bicycles, pedestrians, airplanes, trains, boats, or a combination of these. Factors that transportation engineers look for when designing an operation plan are speed of traffic, how much and what kind of traffic, weather, condition of roads (or railways, sidewalks, ports, etc.), whether or not there is a special event or construction, and countless other factors. Transportation engineers will erect traffic signals, signs, lights, and monitoring devices to conduct and track the flow of traffic.

Research engineers look for ways to improve existing transportation engineering technology and design new tools and methods for the future of the industry. Researchers may study the high accident rate of a certain stretch of highway and come up with ways to minimize hazards. They study the flow and density of traffic and develop a congestion management system to help ease traffic. An important branch of research today is finding ways to reduce pollution resulting from transportation. Other researchers may be involved in designing on-board automobile navigational systems, or a high-speed rapid transit system for a sprawling urban area. There are no boundaries for the transportation research engineer.

WHAT IS IT LIKE TO BE A TRANSPORTATION ENGINEER?

Phil Hartley works as an operations consultant for a busy engineering firm that specializes in urban planning. Phil has numerous responsibilities throughout his workweek (which may include some hours on Saturday). While working on multiple projects in various stages of completion, he must keep track of the needs and challenges of each individual project. In doing so, Phil works an average of fifty to sixty hours a week. "That's mostly in the spring and summer, my busiest time," Phil clarifies. "In the fall and winter I can usually return to a forty-hour workweek."

Phil says that he does not have a typical week. On some mornings he comes into the office and checks first all of his emails and voice mails. He responds to each accordingly. "This can take a few minutes or a few hours, depending on how complex the issues are." He tries to schedule morning meetings with other members of his project team. "The morning is the time to ask questions, give answers, and basically set the agenda for the day or week," Phil explains.

Milestones in transportation history

1854: The internal combustion engine is invented.

Mid-1880s: The first gasoline-powered automobiles are produced.

1903: The Wright brothers make the first successful airplane flight.

1921: The first diesel electric locomotive is introduced.

1927: Charles Lindbergh makes his famous flight over the Atlantic Ocean to Europe.

1940: The first limited-access highway in the United States (the Pennsylvania Turnpike) opens.

1956: Construction of the Interstate Highway System in the United States is authorized by the Federal Aid Highway Act.

1958: The first commercial jet is introduced.

1969: The United States lands a man on the moon.

Info compiled from: Transportation Engineering: An Introduction, by C. Jotin Khisty. Prentice Hall, 1990

On days when Phil is in the office, he generally spends the afternoons working on his computer, figuring out solutions and "action plans" for his projects. Because each transportation operations project is unique, his tasks vary. He works with data he or team members have collected on a certain site. He may conduct mathematical flow rates, analyze speed, flow, and density relationship of a busy stretch of highway, or plan a work safety zone course for construction site traffic, among many other specialized tasks.

As an operations specialist Phil gets to spend a lot of time on-site, away from the office. He sets up computerized monitoring devices, collects data from earlier tests, and surveys project areas. "Sometimes the best way to get a feel for a problem area is to just use it like a regular commuter," Phil explains. He says he has often driven his car at peak rush hour times to help give him a perspective or insight on a trouble area. "Some tests or detailed computer programs don't give you the big picture," he asserts.

Phil works closely with other transportation professionals, local transportation authorities, and planners to learn as much about an area as possible. "Planners are really good at telling us when an idea won't go over well with local officials or neighborhood groups," Phil explains. "They are usually more knowledgeable about local policy issues and regulations than us operations specialists."

Phil also works with pedestrian and bicycle traffic issues. He plans ways to better route pedestrians and cyclists through congested areas. "One of the trickiest issues I've dealt with is finding ways to better assimilate bicyclists into road traffic. There's a lot of tension between the two road users and the potential for an accident is high. It's not as simple as just putting in a bike lane; some streets just can't take one."

Phil has studied areas where motorists and bicyclists have had collisions and he has pinpointed three causes to this problem and made appropriate recommendations to solve the problem. "Visibility was poor due to landscaping," Phil explains, "the speed limit for automobiles was set too high for conditions, and there was no signage up for cyclists to yield or stop." Phil proudly boasts that city officials followed his recommendations and that incidents in the area between bicyclists and automobiles have all but vanished.

Phil's other duties include writing a variety of reports for his senior managers. He writes general progress reports, forecasting reports, and strategy reports. "This is my least favorite part of the job. I don't have time for it, but I recognize its importance," Phil adds. With approaching deadlines Phil has often had to take work home with him at night and on the weekends to complete his reports. Many of these reports entail detailed and complex mathematical problems as well as concise technical writing.

HAVE I GOT WHAT IT TAKES TO BE A TE?

Coordinating hundreds of elements within one large project requires a remarkable ability to keep track of numerous details without losing sight of the big picture. "So many important things in our field are dependent upon the little details," Phil explains. "From my appearance at a public forum after work to my ratio of flow rate calculations, it all has tremendous importance on the end result."

Due to the fact that so many of their projects directly involve the general public, transportation engineers also must zealously guard against errors of any type or scope. A careless mistake could lead to a breakdown in the system or service, potentially posing risks to any number of people. "Public safety is really the underlying purpose of almost every project we do," says Phil.

When mistakes or problems do occur, transportation engineers need to be able to think calmly and clearly in high stress situations. In addition to level-headedness, transportation engineers should have excellent creative problem-solving strategies. "Everyone I know in this field is a troubleshooter," Phil says. "From the engineers-in-training to the most senior engineer, we all love to dissect a problem and see how it can be put back together correctly." Often, that can take a while, so other good qualities to develop are patience and determination.

When solving the various challenges in transportation engineering it helps to have a natural curiosity about systems and methods. Innovation is not the by-product of a static, stagnant mind, so transportation engineers, like other engineers and scientists, should never shrink from asking questions and delving deeper into an area, no matter how obvious or offbeat that investigation may seem at first. Such healthy skepticism and creativity are also useful when working within the constraints of tight budgets, resourcefulness being yet another desirable quality in transportation engineers.

Good communication skills are absolutely essential to the work of a successful transportation engineer. In addition to speaking at presentations within their divisions, they are often called upon to explain at neighborhood meetings why, for example, a stop sign is or isn't necessary at a particular intersection. Being able to eloquently and persuasively present complex scenarios in front of large groups of people helps the public understand the many issues at stake in the decision-making process. Phil often deals directly with clients, neighborhood groups, public officials, and other workers. He must respond to phone calls, write proposals, and go to meetings, while working as a member of a team with his colleagues. "I was in my high school theater club," Phil says. "I'm sure that's what gave me the confidence to stand before a large group of people and defend or present a project without getting nervous." Phil also took a public speaking class in college.

To be a successful transportation engineer, you should:

Have an aptitude for math and science

Be detail-minded and organized

Have good problem-solving skills

Have strong communication skills, both written and oral

Be willing to travel

Have an inquisitive mind

Clearly, transportation engineers must be willing to travel. Typically Phil works on two to three projects at a time, each in different stages of completion. With each project presenting its own unique challenges and requiring a variety of different skills, Phil maintains he never gets bored on the job. "On Monday I was out on a site setting up computerized flow monitors for one project. On Tuesday afternoon I was at a downtown hotel in a meeting with some

Fast Facts

More than half the U.S. population is licensed to drive.

city officials for a new project we're beginning. Wednesday, I was in the office working on some computer simulations of an ITS [Intelligent Transportation System]. And Thursday I was back at the Monday site, checking out a problem with my flow monitors and making other notes on the area. The days usually go by really fast for me."

HOW DO I BECOME A TRANSPORTATION ENGINEER?

EDUCATION

High School

As with most engineering disciplines, it is extremely important to prepare for a college program by taking as many classes in math, science, and computers as possible, including algebra, geometry, calculus, trigonometry, physics, chemistry, and computer science. "The more you know about computers early on, the better off you'll be in any engineering field," Phil stresses. English, speech, and foreign language classes are helpful, too, since engineers need to write and speak well in order to get their ideas across effectively.

Perhaps the next most critical element in preparing for a career in transportation engineering is learning how government works—through study or, better yet, firsthand involvement. Each and every day transportation engineers work with city, state, and national urban planners and government officials to iron out transportation problems. Having a solid grasp of the different branches of government and how they function is an excellent way to prepare for a future career in transportation engineering. Work on a politician's campaign, attend town council meetings, volunteer or apply for part-time work at a municipal division (from traffic court to the mayor's office). Help with the planning and coordination of special civic events. These experiences will not only help you learn about the ways in which different government organizations work together to accomplish goals, but they can help to develop impressive communication skills and valuable networking contacts for the future.

Postsecondary Training

The minimum education requirement for entry-level positions in transportation engineering is a bachelor's degree in a branch of engineering. Most transportation engineers study civil engineering, but other acceptable engineering fields include electrical, mechanical, and chemical. Most of the same principles apply, and any subjects not covered in these majors can be studied on a course-by-course basis, as needed. Many civil engineering programs offer a transportation curriculum. Typical degree programs last four years, although some may last one or two years longer.

Because transportation engineering is an interdisciplinary field, the core transportation curriculum often varies between programs. Generally, first-year students take basic science and math classes, such as statistics, physics, calculus, and introductory courses to the field. Most programs mandate that students also enroll in at least one elective class so they can hone their skills in writing, a foreign language, or political science. English composition classes also are required, as communication with colleagues and clients is essential to success in the field.

Second-year students typically continue to study more advanced science and math classes, but begin taking special courses related to the major, such as civil engineering materials, fundamentals of surveying, basic traffic engineering, and transportation planning.

Courses in advanced transportation engineering classes round out the junior and senior years. Students declare their area of specialization and, depending on that area, take classes such as airport planning and design, geometric design of traffic facilities, public transportation planning, highway design, pavement management systems, river and waterways engineering, open channel hydraulics, work zone safety, and environmental engineering. Computer software programs currently being taught include ArcView and GIS (Geographic Information Systems).

CERTIFICATION OR LICENSING

The general term in engineering for licensed practitioners is professional engineer (PE). While licensing is usually voluntary for many types of engineering, engineers who approve construction plans are required to be licensed by the state in which the construction will occur. Licensing also ensures employers that their employees meet state legal regulations. Phil is not a PE yet, but he plans on taking the final exam for the license within the next two years.

Requirements vary by state, but generally it takes about four to five years to become a licensed PE. Students can begin the process while still in college by taking the Fundamentals of Engineering (FE) exam. This is an eight-hour test that covers the basics of everything from mathematics, electronics, chemistry, and physics, to more advanced knowledge of engineering sciences.

The next step to becoming a PE is acquiring four years of progressive engineering experience. Some states require that engineers obtain their experience under the supervision of a PE. After four years, candidates take the Principles and Practice of Engineering (PPE) exam specific to civil engineering (there's one for almost every other major branch of engineering, too). Those who pass the exam are referred to as professional engineers and are accorded all of the privileges and responsibilities that go along with it.

INTERNSHIPS AND VOLUNTEERSHIPS

Schools requiring an internship usually will set it up with a local or national agency. Internships last between one and two semesters. Students aren't usually paid for their work, but receive credit hours for their participation. While finishing his senior year at Penn State, Phil got an internship through the Pennsylvania Transportation Institute. "I got a good glimpse of the opportunities available in the industry," he says.

WHO WILL HIRE ME?

Phil landed his first job as a transportation engineer for an engineering consulting firm through his college placement office. "There were three job openings in traffic operations in the Chicago area. I applied to all three of them, and received offers from two." Phil actually took the lesser-paying offer because he thought, in the long run, it would lead to more opportunities for advancement. "Already, two years into my job, I'm making much more than the other company offered me. I'm confident I made the right choice."

Phil remembers that there were also recruiters at his campus during his senior year. Many larger companies send recruiters to college campuses to hold interviews with prospective graduates. Some people get job offers before they even graduate. Those who do not receive offers from the recruiters of large corporations still can expect to move into their profession relatively quickly. There are many smaller companies that don't actively recruit on campus but need young, qualified transportation engineers. These positions are posted chiefly at university placement centers, but also in trade journals, in professional association publications, and on the Internet.

Transportation engineers work for local, state, county, or federal governments, or for engineering consulting firms. Other employers are public and private transit authorities, metropolitan planning organizations, academic institutions, transportation equipment manufacturers and suppliers, contracting and construction firms, toll road authorities, port authorities, and parking authorities.

WHO WILL HIRE ME?, CONTINUED

You can apply directly to specific companies or government agencies by sending them your resume along with a cover letter detailing your skills and qualifications and why you would like to work for them.

The Institute of Transportation Engineers publishes a comprehensive list of positions available throughout the United States on their Web page (http://www.ite.org/posavail.htm). You can also find these listings in their monthly magazine, *ITE Journal*. Other engineering-related Web pages provide useful job-hunting information, and allow you to post your resume on their page.

WHERE CAN I GO FROM HERE?

Phil hopes that, with ten more years of traffic operation experience under his belt, he'll be able to start his own traffic consulting business. "There are a lot of small towns, suburbs, even city neighborhoods that are often in need of transportation consulting," Phil says. He is predicting that, as suburban sprawl increases, traffic consultants will be in big demand. Phil cites a recent September 1997 study by the United States Department of Transportation which indicated that many U.S. highways are in urgent need of repair. "A lot of states will be calling on transportation consultants to help them out on this one," Phil says confidently.

With at least three to four years' experience under their belts, transportation engineers move up to positions of more responsibility, such as managing projects. Engineers with good communication skills will be called on to represent their firm or government at neighborhood meetings where plans are introduced and defended. Other engineers change their area of specialty or become even more focused. For example, with the right training and skills, a transportation planner may move up to a high-tech research position.

Many transportation engineers work for government agencies, although it is possible to work as a consultant. For instance, an engineer working at an independent transportation consulting firm may branch off to become self-employed or move up to a more demanding position with a local, state, or federal government agency.

Another career route is to return to academia and enjoy the challenges of teaching others about the intricacies of transportation engineering.

WHAT ARE SOME RELATED JOBS?

The U.S. Department of Labor classifies transportation engineers under the heading Civil Engineering Occupations. Some related jobs include airport engineers, civil engineers, civil engineering technicians, drafters, forest engineers, highway-administrative engineers, railroad engineers, drainage-design coordinators, structural engineers, sanitary engineers, waste-management engineers, hydraulic engineers, and irrigation engineers.

Related Jobs

Airport engineers

Civil engineering technicians

Civil engineers

Drafters

Drainage-design coordinators

Forest engineers

Highway-administrative engineers

Hydraulic engineers

Irrigation engineers

Railroad engineers

Sanitary engineers

Structural engineers

Waste-management engineers

WHAT ARE THE SALARY RANGES?

A recent graduate salary survey conducted by the Institute of Transportation Engineers concluded that the average annual salary for entry level transportation engineers with bachelor's degrees is $34,865. Those with their master's degree earn on average $37,470, and Ph.D.s average $41,662. Salaries may vary with type of work, whether in planning, design, operations, or research, and depending on where you work, either for a government agency, private practice engineering firm, or as a consultant. The federal government pays the highest, followed by private practice.

Transportation engineers in mid-career with at least sixteen years experience earn on average $66,000. Senior engineers with over twenty-five years experience can earn between $78,000 and $93,000, according to a 1995 survey by the Institute of Transportation Engineers.

Transportation engineers can usually expect a good benefits package, including paid sick and holiday time, two weeks vacation, personal time, medical coverage, stock options, 401K plans, and other perks, depending on the company and industry. "I received a pretty nice profit sharing bonus last year," Phil says. "Business is looking really good in the industry, so I can expect another nice one this year."

WHAT IS THE JOB OUTLOOK?

Transportation engineers should anticipate a promising future because transportation engineering has traditionally been a subfield of civil engineering; a proper transportation curriculum in many university programs was not available, or was at best minimal, for many years. This caused a problematic shortage of qualified transportation engineers. Demand has been strong for transportation-trained engineers, and for civil engineering programs to offer a more comprehensive transportation curriculum. This demand has been met in recent years and more civil engineers are choosing to specialize in transportation. Still, there is a need for transportation professionals to take on new positions as well as replace an estimated one-third of transportation engineers who are predicted to retire within the next five years.

Growth in the industry is expected to be as fast as the average through the year 2005. The continued development of suburban areas is expected to provide many opportunities for transportation professionals, as many of these suburbs approach city-size. Conversely, major cities that are losing large percentages of their population to the suburbs are planning many urban renewal projects to lure people back. They will rely heavily on transportation engineers in their cause to make downtowns and surrounding neighborhoods more people- and car-friendly through expanded public transportation networks, better streets and traffic systems, and improved pedestrian and bicycle transportation routes.

Because many of the major transportation jobs are funded by the federal government, employment is tied to government activity and the economy. Some transportation hiring and new projects fall prey to budget cuts and politics. However, recent federal legislation has increased the transportation pool according to the Institute of Transportation Engineers.

The growth of our nation and the world demands a reliable, efficient, and safe infrastructure. Cities and suburbs change. Populations shift. Transportation facilities wear down or become inefficient. All of these certainties will require the skills of transportation engineers.

What Can I Do Right Now?

engineering

Get Involved: A Directory of Camps, Programs, Competitions, Etc.

Now that you've read about some of the different careers available in engineering, you may be anxious to experience this line of work for yourself, to find out what it's *really* like. Or perhaps you already feel certain that this is the career path for you and want to get started on it right away. Whichever is the case, this section is for you! There are plenty of things you can do right now to learn about engineering careers while gaining valuable experience. Just as important, you'll get to meet new friends and see new places, too.

In the following pages you will find programs designed to pique your interest in engineering and start preparing you for a career. You already know that this field is complex and highly technical, and that to work in it you need a solid education. Since the first step toward an engineering career will be gaining that education, we've found more than thirty programs that will start you on your way. Some are special introductory sessions, others are actual college courses—one of them may be right for you. Take time to read over the listings and see how each compares to your situation: how committed you are to engineering, how much of your money and free time you're willing to devote to it, and how the program will help you after high school. These listings are divided into categories, with the type of program printed right after its name or the name of the sponsoring organization.

THE CATEGORIES

Camps

When you see an activity that is classified as a camp, don't automatically start packing your tent and mosquito repellent. Where academic study is involved, the term "camp" often simply means a residential program including both educational and recreational activities. It's sometimes hard to differentiate between such camps and other study programs, but if the sponsoring organization calls it a camp, so do we!

College Courses/Summer Study

These terms are linked because most college courses offered to students your age must take place in the summer, when you are out of school. At the same time, many summer study programs are sponsored by colleges and universities that want to attract future students and give them a head start in higher education. Summer study of almost any type is a good idea because it keeps your mind and your study skills sharp over the long vacation. Summer study at a college offers any number of additional benefits, including giving you the tools to make a well-informed decision about your future academic career. Study options, including some impressive college and university programs, account for most of the listings in this section—primarily because higher education is so crucial to every engineering career.

Competitions

Competitions are fairly self-explanatory, but you should know that there are only a few in this book for two reasons: one, a number of engineering competitions are part of general science contests too numerous to mention, and two, many engineering competitions are at the local and regional levels and so again are impractical to list here. What this means, however, is that if you are interested in entering a competition, you shouldn't have much trouble finding one yourself. Your guidance counselor or math and science teachers can help you start searching in your area.

Conferences

Conferences for high school students are usually difficult to track down because most are for professionals in the field who gather to share new information and ideas with each other. Don't be discouraged, though. A number of professional organizations with student branches invite those student members to their conferences and plan special events for them. Some student branches even run their own conferences; check the directory of student organizations at the end of this section for possible leads. This is an option worth pursuing because conferences focus on some of the most current information available and also give you the chance to meet professionals who can answer your questions and even offer advice.

Employment and Internship Opportunities

As you may already know from experience, employment opportunities for teenagers can be very limited. This is particularly true in engineering, which requires workers with bachelor's and even graduate degrees. Even internships are most often reserved for college students who have completed at least one or two years of study in the field. Still, if you're very determined to find an internship or paid position in engineering, there may be ways to find one. See the "Do It Yourself" section in this book for some suggestions.

Field Experience

This is something of a catch-all category for activities that don't exactly fit the other descriptions. But anything called a field experience in this book is always a good opportunity to get out and explore the work of engineering professionals.

Membership

When an organization is in this category, it simply means that you are welcome to pay your dues and become a card-carrying member. Formally joining any organization brings the benefits of meeting others who share your interests, finding opportunities to get involved, and keeping up with current events. Depending on how active you are, the contacts you make and the experiences you gain may help when the time comes to apply to colleges or look for a job.

In some organizations, you pay a special student rate and receive benefits similar to regular members. Many organizations, however, are now starting student branches with their own benefits and publications. As in any field, make sure you understand exactly what the benefits of membership are *before* you join.

Finally, don't let membership dues discourage you from making contact with these organizations. Some charge dues as low as $2 because they know that students are perpetually short of funds. When the annual dues are higher, think of the money as an investment in your future and then consider if it is too much to pay.

Incidentally . . .
We have included in these pages a few listings for localized programs and activities that are actually connected to larger movements or organizations. For example, some programs listed here are run by university branches of the Society of Women Engineers (SWE)—a national organization. Others are part of the Minority Introduction to Engineering (MITE) program. If you're interested in SWE or MITE, check with the colleges and universities that interest you to see if they have a branch of their own.

PROGRAM DESCRIPTIONS

Once you've started to look at the individual listings themselves, you'll find that they contain a lot of information. Naturally, there is a general description of each program, but wherever possible we also have included the following details.

Application Information
Each listing notes how far in advance you'll need to apply for the program or position, but the simple rule is to apply as far in advance as possible. This ensures that you won't miss out on a great opportunity simply because other people got there ahead of you. It also means that you will get a timely decision on your application, so if you are not accepted, you'll still have some time to apply elsewhere. As for the things that make up your application—essays, recommendations, etc.—we've tried to tell you what's involved, but be sure to contact the program about specific requirements before you submit anything.

Background Information
This includes such information as the date the program was established, the name of the organization that is sponsoring it financially, and the faculty and staff who will be there for you. This can help you—and your family—gauge the quality and reliability of the program.

Classes and Activities
Classes and activities change from year to year, depending on popularity, availability of instructors, and many other factors. Nevertheless, colleges and universities quite consistently offer the same or similar classes, even in their summer sessions. Courses like "Introduction to Mechanical Engineering" and "Physics 101," for example, are simply indispensible. So you can look through the listings and see which programs offer foundational courses like these and which offer courses on more variable topics. As for activities, we note when you

have access to recreational facilities on campus, and it's usually a given that special social and cultural activities will be arranged for most programs.

Contact Information

Wherever possible, we have given the *title* of the person whom you should contact instead of the *name* because people change jobs so frequently. If no title is given and you are telephoning an organization, simply tell the person who answers the phone the name of the program that interests you and he or she will forward your call. If you are writing, include the line "Attention: Summer Study Program" (or whatever is appropriate after "Attention") somewhere on the envelope. This will help to ensure that your letter goes to the person in charge of that program.

Credit

Where academic programs are concerned, we sometimes note that high school or college credit is available to those who have completed them. This means that the program can count toward your high school diploma or a future college degree just like a regular course. Obviously, this can be very useful, but it's important to note that rules about accepting such credit vary from school to school. Before you commit to a program offering high school credit, check with your guidance counselor to see if it is acceptable to your school. As for programs offering college credit, check with your chosen college (if you have one) to see if they will accept it.

Eligibility and Qualifications

The main eligibility requirement to be concerned about is age or grade in school. A term frequently used in relation to grade level is "rising," as in "rising senior": someone who will be a senior when the next school year begins. This is especially important where summer programs are concerned. A number of university-based programs make admissions decisions partly in consideration of GPA, class rank, and standardized test scores. This is mentioned in the listings, but you must contact the program for specific numbers. If you are worried that your GPA or your ACT scores, for example, aren't good enough, don't let them stop you from applying to programs that consider such things in the admissions process. Often, a fine essay or even an example of your dedication and eagerness can compensate for statistical weaknesses.

Facilities

We tell you where you'll be living, studying, eating, and having fun during these programs, but there isn't enough room to go into all the details. Some of those details can be important: what is and isn't accessible for people with disabilities, whether the site of a summer program has air-conditioning, and how modern the laboratory and computer equipment are. You can expect most program brochures and application materials to address these concerns, but if you still have questions about the facilities, just call the program's administration and ask.

Financial Details

While a few of the programs listed here are fully underwritten by collegiate and corporate sponsors, most of them rely on you for at least some of their funding. Prices given here are rounded up from 1997 figures, but you should bear in mind that prices rise slightly almost every year. You and your parents must take costs into consideration when choosing a program. We always try to note where financial aid is available, but really, most programs will do their best to ensure that a shortage of funds does not prevent you from taking part.

Minorities

In the not-so-distant past, engineering was a field populated almost solely by Caucasian males—that is most certainly not the case today. Still, colleges and universities are working to promote even more diversity in the field, so there are any number of programs encouraging applicants of every gender, race, and ethnic background. However, when an engineering department bills its summer session as a "minority program," you must determine just who is a minority. If you have an Asian heritage, you are a minority in the U.S. population, but not necessarily in engineering: Asian-Americans are so well represented in the field that the MITE program, among others, does not include them. In short, if you are interested in minority programs, make sure you know how they define the term.

Residential vs. Commuter Options

Simply put, some programs prefer that participating students live with other participants and staff members, others do not, and still others leave the decision entirely to the students themselves. As a rule, residential programs are suitable for young people who live out of town or even out of state, as well as for local residents. They generally provide a better overview of college life than programs in which you're only on campus for a few hours a day, and they're a way to test how well you cope with living away from home. Commuter programs may be viable only if you live near the program site or if you can stay with relatives who do. Bear in mind that for residential programs especially, the travel between your home and the location of the activity is almost always your responsibility and can significantly increase the cost of participation.

FINALLY . . .

Ultimately, there are three important things to bear in mind concerning all of the programs listed in this volume. The first is that things change. Staff members come and go, funding is added or withdrawn, supply and demand determine which programs continue and which terminate. Dates, times, and costs vary widely because of a number of factors. Because of this, the information we give you, although as current and detailed as possible, is just not enough on which to base your final decision. If you are interested in a program, you simply must write, call, fax, or email the organization concerned to get the latest and most complete information available. This has the added benefit of putting you in touch with someone who can deal with your individual questions and problems.

Another important point to keep in mind when considering these programs is that the people who run them provided the information printed here. The editors of this book haven't attended the programs and don't endorse them: we simply give you the information with which to begin your own research. And after all, we can't pass judgment because you're the only one who can decide which programs are right for you.

The final thing to bear in mind is that the programs listed here are just the tip of the iceberg. No book can possibly cover all of the opportunities that are available to you—partly because they are so numerous and are constantly coming and going, but partly because some are waiting to be discovered. For instance, you may be very interested in taking a college course but don't see the college that interests you in the listings. Call their Admissions Office! Even if they don't have a special program for high school students, they might be able to make some kind of arrangements for you to visit or sit in on a class. Use the ideas behind these listings and take the initiative to turn them into opportunities!

THE PROGRAMS

AMERICAN CHEMICAL SOCIETY
Field Experience, Membership

The American Chemical Society (ACS), formed in the 1930s, runs a number of educational and career-related programs for high school students interested in chemistry and chemical engineering. One of their projects links disadvantaged students with chemical researchers in industry and government, providing hands-on experience and extra encouragement. The ACS also publishes a magazine for high school chemistry students, *ChemMatters,* and even has high school curricula of its own. If you're interested in the chemical aspects of engineering, get in touch with the American Chemical Society. If you're visiting their Web site, look under "educational programs."

American Chemical Society
1155 16th Street, NW
Washington, DC 20036
Tel: 202-872-4600
Web: http://www.acs.org

AMERICAN INDIAN SCIENCE & ENGINEERING SOCIETY

Conference, Membership

The American Indian Science & Engineering Society (AISES) is a nonprofit organization founded in 1977 to encourage young American Indians to pursue careers in science, technology, and engineering. AISES welcomes students from kindergarten to twelfth grade (whom they refer to as "pre-college students") as members; dues are only $2 per year. Pre-college student members can participate in many special programs, including the National American Indian Science and Engineering Fair, and attend the annual AISES National Conference, which features High School Day and a career fair. AISES also has a number of programs designed for entire classes and their teachers, so talk to your science or math teacher about getting involved with the American Indian Science & Engineering Society. You can also contact them yourself or visit their Web site for more information.

American Indian Science & Engineering Society
5661 Airport Boulevard
Boulder, CO 80301-2339
Tel: 303-939-0023
Web: http://www.colorado.edu/AISES
Email: aisespc@spot.colorado.edu

BUSINESS, ENGINEERING, SCIENCE & TECHNOLOGY (BEST) SCHOLARS PROGRAM

College Course/Summer Study

The Pennsylvania State University (Penn State) sponsors the Business, Engineering, Science & Technology (BEST) Scholars Program each summer in conjunction with the Eastman Kodak Company. The program is for rising seniors who are residents of Pennsylvania or Rochester, New York, and are members of an underrepresented minority group: Native American, African-American, Latino/Hispanic American. BEST participants learn about business, engineering, and scientific educational opportunities at Penn State and career opportunities at Eastman Kodak. Students live at Penn State's University Park campus for four weeks beginning in early July. While on campus you take enrichment classes in mathematics, communications, and computer science and also tour the relevant academic facilities. There are workshops on college admission and financial aid, weekly seminars on career options by professionals from Eastman Kodak, and a trip to Kodak's facilities in Rochester, New York. Many special social activities are also planned for the participants. After successfully completing the program, you receive a certificate, a $400 stipend, and a Kodak mentor to help and advise you for the next five years. There is no cost for BEST; students must, however, provide their own transportation to and from campus. This is a unique opportunity for those who are serious about a business, engineering, or science career, so admission to the program is competitive. You must have a B+ GPA or higher and provide an essay, transcript, and teacher recommendation along with your application form; applications are due at the beginning of April. For a form and further information, contact the Program Director.

■ **Business, Engineering, Science & Technology (BEST) Scholars Program**
The Pennsylvania State University
241 Hammond Building
University Park, PA 16802
Tel: 800-848-9223

THE CENTER FOR EXCELLENCE IN EDUCATION

Field Experience

The goal of the Center for Excellence in Education (CEE) is to nurture future leaders in science, technology, and business. And it won't cost you a dime: all of CEE's programs are absolutely free. Since 1984, the CEE has sponsored the Research Science Institute, a six-week residential summer program held at Massachusetts Institute of Technology. Seventy high school students with scientific and technological promise are chosen from a field of over seven hundred applicants to participate in the program, conducting projects with scientists and researchers. You can read more about specific research projects on-line.

Another CEE program, the Role Models and Leaders Project, is currently underway in Washington, DC; Los Angeles; and Chicago. It targets students who have the ability but not the economic and social advantages to excel in science and math, but they've got to be willing to sacrifice Saturday mornings for a couple of years. Participants attend twenty Saturday morning sessions during their junior and senior years of high school, when mentors lead discussions about scientific concepts, career opportunities, the college selection and application process, and scholarships and financial aid. When the students go off to college, mentors continue to provide moral support and tutorial assistance during the transition. If you think you fit the bill for either of these programs, simply request an application form. If the Role Models and Leaders Project is not offered in your city but you think there's a need for it, be sure to express your interest.

■ **The Center for Excellence in Education**
7710 Old Springhouse Road, Suite 100
McLean, VA 22102
Tel: 703-448-9062
Web: http://rsi.cee.org/index.html
Email: cee@pop.erols.com

CHALLENGE PROGRAM AT ST. VINCENT COLLEGE

College Course/Summer Study

The Challenge Program, offered by St. Vincent College, is just what its name implies. Challenge gives gifted, creative, and talented students in grades nine through twelve the opportunity to explore new and stimulating subjects that most high schools just can't cover. If you qualify for this program and are highly motivated, you can spend one week in July on the campus of St. Vincent College taking courses like robotics and aviation. Should you choose, you may live on campus, meeting and socializing with other students who share your ambitions and interests. Resident students pay a total of about $475 for the week while commuters pay closer to $325. For more information about Challenge and details of this year's course offerings, contact the Program Coordinator. There is a similar Challenge program for students in the sixth through ninth grades, usually held one week before the high school session.

■ **Challenge Program at St. Vincent College**
300 Fraser Purchase Road
Latrobe, PA 15650
Tel: 412-537-4569

CHEMISTRY IS FUN! CAMP

Camp

The Center for Chemical Education at Miami University of Ohio sponsors a series of Terrific Science Camps each summer. Rising ninth- and tenth-graders are welcome to attend the Chemistry Is Fun! Camp. It lasts three hours every day for one week during the month of July. As a participant, you explore the fascinating world of chemistry and discuss recent events and breakthroughs in the field. Best of all, you conduct your own chemical experiments in Miami University's chem lab. If you're considering a career in chemical engineering, this is obviously a good opportunity to start exploring. But Chemistry Is Fun! is also a fine option if you're thinking vaguely of a career in science or if you want to prepare a little for a high school chemistry class. The cost for this commuter camp is only about $75, and free tuition is available to students truly in need of financial aid. Admission is made on a first come, first served basis, so you should apply well before the deadline in late April. Please note that while most of the Terrific Science Camps are held on Miami University's Middletown campus, the Chemistry Is Fun! Camp is usually held on the campus in Oxford, Ohio. For an application form and more information on this year's camp, call or write the Terrific Science Camps at the Center for Chemical Education.

■ **Chemistry Is Fun! Camp**
Miami University Middletown
4200 East University Boulevard
Middletown, OH 45042
Tel: 513-727-3269

COLLEGE AND CAREERS PROGRAM

College Course/Summer Study

The Rochester Institute of Technology (RIT) offers its College and Careers Program for rising seniors who want to experience college life and explore career options in engineering. The program, in existence since 1990, allows you to spend a Friday and Saturday on campus, living in the dorms and attending four sessions on the career areas of your choice. Each year, some sessions focus on the liberal arts and sciences, but between fifteen and twenty sessions focus on engineering and technology. Topics vary, but generally include introductions to civil, electrical, mechanical, and manufacturing engineering, among others. In each session, participants work with RIT students and faculty to gain some hands-on experience in the topic area. This residential program is held twice each summer, usually once in mid-July and again in early August. The registration deadline is one week before the start of the program, but space is limited and students are accepted on a first come, first served basis. The cost for the College and Careers Program is $45, although students may choose not to spend Friday night on campus and pay only $30. For further information about the program and specific sessions on offer, contact the RIT admissions office.

College and Careers Program
Rochester Institute of Technology
Office of Admissions, 60 Lomb Memorial Drive
Rochester, NY 14623-5604
Tel: 716-475-6635

CORNELL UNIVERSITY SUMMER COLLEGE

College Course/Summer Study

As part of its Summer College for High School Students, Cornell University offers an Exploration in Engineering for students who have completed their junior or senior years. The Summer College session runs for six weeks from late June until early August. It is largely a residential program designed to acquaint you with all aspects of college life. The Exploration in Engineering seminar is one of several such seminars offered by Cornell to allow students to survey various disciplines within the field and speak with working professionals. The seminar meets several times per week and includes laboratory projects and field trips. In addition, Summer College participants take two college-level courses of their own choosing, one of which should be in computer science or mathematics to complement the engineering focus. You must bear in mind that these are regular undergraduate courses condensed into a very short time span, so they are especially challenging and demanding. Besides the course material, you will learn time-management and study skills to prepare you for a program of undergraduate study.

Cornell University awards letter grades and full undergraduate credit for the two courses you complete, but none for the Exploration. Residents live and eat on campus, and enjoy access to the university's recreational facilities and special activities. Academic fees total around $3,400, while housing, food, and recreation fees amount to an additional $1,600. Books, travel, and an application fee are extra. A very limited amount of financial aid is available. Applications are due in early May, although Cornell advises that you submit them well in advance of the deadline; those applying for financial aid must submit their applications by early April. Further information and details of the application procedure are available from the Summer College office.

■ **Cornell University Summer College for High School Students**
School of Continuing Education and Summer Sessions
B20 Day Hall
Ithaca, NY 14853-2801
Tel: 607-255-6203

ENGINEERING 2003

College Course/Summer Study

The Catholic University of America (CUA) offers rising seniors the chance to participate in Engineering 2003 (the date changes yearly, to reflect the date when participants will graduate from a four-year university course). Students in the program explore the opportunities available to them in this field and even consider an undergraduate engineering degree as a solid foundation for graduate studies in law, medicine, and business. During the week-long program, held each July since 1988, participants take part in two research projects from a selection of subjects like Robotics and Artificial Intelligence, Computer Graphics, and Surveying and Construction. As a participant, you also take an active role in discussions with CUA engineering faculty and students about such topics as engineering curricula and careers, and professional ethics. There is also time for you to take advantage of CUA's many recreational facilities and to explore campus life while living in the dormitories. Applicants to Engineering 2003 must submit a completed form, transcript, and one letter of recommendation; participants are selected primarily on the basis of their achievements in science and mathematics. Upon acceptance, students must pay a $200 registration fee, which is the only cost of the program except for transportation to and from Engineering 2003. For an application and further information, contact the Dean's Office at the School of Engineering.

■ **Engineering 2003**
The Catholic University of America, School of Engineering
Room 102, Pangborn Hall
Washington, DC 20077-5216
Tel: 202-319-5160
Web: http://www.ee.cua.edu
Email: cua-engineer@cua.edu

ENGINEERING SUMMER ACADEMY AT THE UNIVERSITY OF TULSA

College Course/Summer Study

The University of Tulsa hosts the Engineering Summer Academy in conjunction with the Oklahoma Board of Regents each June. The month-long program (which includes a break for Father's Day) is for rising freshmen and sophomores who are residents of Oklahoma. Participants live on campus and take part in such activities as a design project, lab experiments, site visits, computer work, and shadowing a professional engineer for a day. Students also explore career options and the philosophy and ethics of engineering. The University of Tulsa hosts four follow-up sessions during the school year where students can continue their activities and explore further. The Engineering Summer Academy is fully funded by the Oklahoma Board of Regents, so students need pay only for travel to and from the campus and incidental expenses. Applicants with a good academic background, ability to succeed in the program, and who are likely to pursue a career in engineering have the best chance of being selected. Students from minority groups generally underrepresented in the field of engineering (females, ethnic minorities, the disabled) are strongly encouraged to apply, but the program is open to everyone who meets the basic age and residency requirements. The application form and one recommendation from a teacher are due before the beginning of March. Semifinalists are interviewed in Tulsa at the end of March and notified of the final decisions shortly thereafter. For further information and an application form, contact the Engineering Summer Academy.

> **Engineering Summer Academy at the University of Tulsa**
> Department of Chemical Engineering
> 600 South College Avenue
> Tulsa, OK 74104
> Tel: 918-631-2226

ENGINEERING SUMMER INSTITUTE FOR HIGH SCHOOL STUDENTS

College Course/Summer Study

The College of Engineering at Southern University sponsors the Engineering Summer Institute for High School Students (ESIHSS) for those who are rising to the ninth through twelfth grades. In operation since 1974, the ESIHSS is a four-week residential program held on the Southern University campus to encourage young people to pursue careers in engineering, math, and the sciences. It generally runs from early June to early July. Participants are divided into two groups: ninth- and tenth-graders, and eleventh- and twelfth-graders. At both levels, students take part in classroom and laboratory exercises, study science ethics, and interact with people who are role models in the field. On Fridays, you visit such companies as Dow Chemical and the EXXON Corporation to see engineering and scientific principles applied to industry. ESIHSS participants also benefit from specialized career counseling. Interested students must submit an application form, essay, transcript including standardized test scores, and a letter of recommendation from a recent math or science teacher by the beginning of April. Because there are only forty available places and applications are accepted from across the country, competition is strong and only the most promising applicants (with a GPA of at least 3.0) are accepted. Once accepted, students must pay a $300 registration fee and cover their own travel and personal expenses. All other costs, such as tuition, and room and board, will be absorbed by the university and its supporters. For further information and an application form, contact the ESI Director. Southern University also sponsors slightly different versions of the Engineering Summer Institute for Elementary School Students (ESIESS) and for Middle School Students (ESIMSS). These programs are open only to students from the Baton Rouge Metropolitan Statistical Area. Contact the ESI Director for further information.

Engineering Summer Institute for High School Students
College of Engineering
Southern University
Baton Rouge, LA 70813
Tel: 504-771-3798

ENGINEERING SUMMER RESIDENCY PROGRAM (ESRP)

College Course/Summer Study

The University of California, Davis (UCD), welcomes rising juniors and seniors from population groups that traditionally have been underrepresented in engineering to its Engineering Summer Residency Program (ESRP). Running for one week in late June, ESRP introduces students to the field of engineering and the many different career opportunities within it. Participants get a genuine taste of college life, including a daily schedule of lectures, laboratories, and demonstrations on various aspects of engineering and on computer science. You live in university dormitories, and spend your time with UCD students and staff in social as well as academic activities. There is no cost for this program except for travel to and from the Davis campus and personal expenses. Applications must be submitted by the first Friday in April; priority is given to applicants with strong preparation in math and science.

Engineering Summer Residency Program (ESRP)
Special Programs in Engineering
University of California, Davis
Dean's Office, College of Engineering
1050 E Engineering II
Davis, CA 95616
Tel: 916-752-3316

FIRST

Competition

The nonprofit organization FIRST—For Inspiration and Recognition of Science and Technology—wants to generate an interest in science and engineering in high school students. And judging by their annual robot-building contest, interest is not a problem. Each year the rules of the competition are different, and details are kept top-secret until a kick-off workshop. Then, for six weeks, students team up with engineers from corporations and universities to brainstorm, design, construct, and test their robots. The 1996 competition involved building "robo-gladiators," using materials donated by numerous corporate sponsors. If you want to know exactly what a robo-gladiator is, pay a visit to FIRST's Web site for photographs and a complete description. You can even download a video from the final competition, which was held at Epcot Center in Florida. Of course, you can also contact FIRST by mail, phone, and fax to get more information.

> **FIRST**
> 200 Bedford Street
> Manchester, NH 03101
> Tel: 800-871-8326
> Web: http://usfirst.mv.com/

FUTURE SCIENTISTS AND ENGINEERS OF AMERICA

Membership

Future Scientists and Engineers of America (FSEA) is a nonprofit organization on the order of Future Farmers of America, working to encourage early participation in a chosen career field. The FSEA encourages young people in kindergarten through twelfth grade to found student chapters in their schools. Each chapter needs a sponsor from the local community, a teacher to act as an advisor, a parent coordinator, two mentors from the local scientific or engineering community, and up to twenty-five student members. Naturally, with these people involved, the students in each FSEA chapter learn about the professional world and gain hands-on experience in science and engineering. If you'd like more information about the FSEA or about starting a chapter, write to them or look up their Web site.

> **Future Scientists and Engineers of America**
> PO Box 9577
> Anaheim, CA 92812
> Web: http://www2.fsea.org/FSEA

HILA SCIENCE CAMP

Camp

Hila Science Camp offers both residential and day camps in the Ottawa Valley, close to Beachburg, Ontario. Hila, established in 1984, features a variety of programs with an emphasis on science and technology. You choose a major program as your focus; options include engineering and technology, rocketry, and computer science. You can build an advanced model rocket, construct and fly a balsa model aircraft, and construct electronic circuits using circuit boards, solder, and a soldering iron. The computer science program shows you how to create digital computer images, set up your own Web page, and write computer programs. In addition to your major program, you can collect specimens, go on archeological digs, hunt for fossils, and gaze at the stars through Hila's eight-foot reflecting telescope. Residential camp lasts one week, from Sunday afternoon through the following Friday. Campers sleep in large dorm-style tents, and Hila facilities include a dining hall, basketball court, rocketry and aircraft shop, computer room, and a flying field. Day camp runs Monday through Friday, 9:00 AM to 5:00 PM. Hila offers six camp sessions, beginning in July, and all programs are offered in all sessions. A week of residential camp at Hila costs about $460; day camp is $260. Kits and equipment for various programs cost extra. Supplies for the engineering and technology program, for example, cost an extra $70. Each program enrolls a maximum of thirty-four students on a first come, first served basis. If you live in the Ottawa Valley, some scholarships are available. Hila is happy to provide you with more details on all matters.

> ■ **Hila Science Camp**
> RR #2
> Pembroke, Ontario, Canada K8A 6W3
> Tel: 613-582-3632
> Web: http://fox.nstn.ca/~hila/index.html
> Email: hila@fox.nstn.ca

IDAHO JEMS SUMMER WORKSHOP

College Course/Summer Study

The University of Idaho College of Engineering sponsors Idaho JEMS (Junior Engineering Math & Science program) each summer for rising seniors. During Idaho JEMS, first introduced in 1967, students live on the university campus for two weeks and take classes with College of Engineering professors. Such classes may include engineering design, human factors, computer programming, and engineering problem-solving. Successful completion of the classes leads to college engineering credits from the University of Idaho. In addition to course work, participants explore engineering through lab exercises, field trips, and guest speakers. You can experience college life to the fullest while living and dining on campus by using the university's recreational facilities and touring its colleges. There are also special barbecues and dances to promote an active social life during the Idaho JEMS program. Applicants must have at least a 3.0 GPA and three years of mathematics; you must submit a high school transcript, brief resume, and essay along with your application form. The cost of the program is about $400 including room and board, and some financial aid is available. Female and minority students are especially encouraged to apply, but the program is open to all qualified applicants. For more information and an application, contact the Program Director.

Idaho JEMS Summer Workshop
College of Engineering
University of Idaho
Moscow, ID 83844-1011
Tel: 208-885-7303

INTERNATIONAL BRIDGE BUILDING COMPETITION

Competition

The International Bridge Building Competition is an annual opportunity to test your ability to construct a model bridge based on sound engineering principles and a set of specifications. Once your bridge is built to specification, it must withstand a certain amount of weight and be fully functional. Bridges are judged for efficiency, and the most efficient bridge wins. The contest, which originated at Illinois Institute of Technology in 1977, is held regionally across America. Each region sends its two top winners to the International Bridge Building Competition, held in a different location every year. There is no cost to enter the competition, although certain regions may charge a small amount for a bridge kit containing everything you need to build your bridge. You must enter the competition through your school, and schools must be registered with a region. Some schools and regions allow for teams, but only one person can be declared the winner at the international level. First place winners can take advantage of a four-year scholarship to the Illinois Institute of Technology. Other prizes vary from year to year. All participants receive medals and a trophy for their school. For details and a list of regions, contact the International Bridge Building Competition at the Illinois Institute of Technology.

International Bridge Building Competition
Illinois Institute of Technology
Chicago, IL 60616
Tel: 773-567-3498
Web: http://www.iit.edu/~hsbridge/
Email: segre@itt.edu

JUNIOR ENGINEERING TECHNICAL SOCIETY (JETS)

Competition, Field Experience

The Junior Engineering Technical Society (JETS) offers several different opportunities for young people in grades nine through twelve to test and strengthen their aptitude for engineering before making college and career decisions. One of their most interesting and direct means of testing aptitude is administrating the National Engineering Aptitude Search+ (NEAS+) exam. The NEAS+ is a self-assessment of your thinking, reasoning, and understanding processes, which you can take at any point during your high school career. This not only reveals whether your skills are generally suited to an engineering career, but also what your weaknesses are so that you can work to improve them before starting college. To take this exam, you need only request the NEAS+ kit (which includes career guidance materials) from JETS; the cost is about $15.

If some of your friends share your interest in engineering, you may want to participate in JETS' Tests of Engineering Aptitude, Mathematics, and Science (TEAMS). The TEAMS program is a one-day assessment of your ability to work with others on engineering problems; results determine if your group is ranked at the local, state, or national levels. High school groups can also participate in JETS' National Engineering Design Challenge (NEDC), a year-long program and competition also leading to rankings at the local, state, and national levels. In the NEDC, you concentrate on problem-solving and team-building exercises as you consider real challenges from the world of engineering. For both the TEAMS and NEDC programs, your group needs a teacher to serve as your coach, so speak to your math and science teachers and ask them to contact JETS for detailed information.

Junior Engineering Technical Society (JETS)
1420 King Street
Suite 405
Alexandria, VA 22314-2794
Tel: 703-548-5387
Web: http://www.asee.org/jets
Email: jets@nae.edu

JUNIOR SCIENCE AND HUMANITIES SYMPOSIUM (JSHS)

Conference

The Junior Science and Humanities Symposium encourages high school students (grades nine through twelve) who are gifted in engineering, mathematics, and the sciences to develop their analytical and creative skills. There are nearly fifty symposia held at locations all around the United States—including Georgetown University, the University of Toledo, and Seattle Pacific University—so that each year some ten thousand students are able to participate. Funded by the U.S. Army Research Office since its inception in 1958 (and by the Office of Naval Research and the Air Force Office of Scientific Research since 1995), the JSHS has little to do with the military and everything to do with research. At each individual symposium, researchers and educators from various universities and laboratories meet with the high school students (and some of their teachers) to study new scientific findings, pursue their own interests in the lab, and discuss and debate relevant issues. Participants learn how scientific and engineering research can be used to benefit humanity, and they are strongly encouraged to pursue such research in college and as a career. To provide further encouragement, one attendee at each symposium will win a scholarship of about $4,000 and the chance to present his or her own research at the national Junior Science and Humanities Symposium. Four other attendees from each regional JSHS win all expense paid trips to the national symposium, where the top research students can win additional scholarships worth up to $20,000 and trips to the prestigious London International Youth Sciences Forum. For information about the symposium in your region and on eligibility requirements, contact the national Junior Science and Humanities Symposium.

■ Junior Science and Humanities Symposium (JSHS)
Academy of Applied Science
PO Box 2934
Concord, NH 03302-2934
Tel: 603-228-4520

LEONARDO DA VINCI COMPETITION

Competition

The Leonardo da Vinci Competition is a written test on engineering-oriented problems covering such topics as physics, chemistry, calculus, algebra, and geometry. It is available to Canadian high school students in their senior year. If you're applying for admission to an engineering program at the University of Toronto, you must enter the competition. Or, if you just like science and math and want to determine if you have an aptitude for engineering, this test is also for you. Prizes include a $2,000 scholarship for each of the top fifteen candidates who apply for a full-time first-year engineering program at the University of Toronto the September following the competition. In addition, the first place winner is awarded $500, second place winner receives $400, third place $300, fourth place $200, and sixth through tenth place winners are awarded $100 each.

The Leonardo da Vinci test is sent out to Canadian high schools in January and administered in April. The cost to take the test is $7 per student, and a teacher must volunteer to coordinate and administer the exam. To see if your school is on the mailing list, you can check the da Vinci Web site. Your teacher can add your school to the mailing list right at the site. For more information, call, write, or email the da Vinci Competition Office.

■**Leonardo da Vinci Competition**
University of Toronto—Engineering
35 St. George Street, Room 170
Toronto, Ontario, Canada M5S 1A4
Tel: 416-978-5034
Web: http://www.ecf.toronto.edu/apsc/davinci
Email: davinci@ecf.utoronto.ca

THE MAKING OF AN ENGINEER

College Course/Summer Study

The University of Denver invites rising sophomores, juniors, and seniors to apply for its Making of an Engineer course, which runs for three weeks in June. As a part of the university's larger Early Experience Program, the course encourages students already interested in science and technology to consider putting those interests to work in an engineering career. All participants attend lectures and laboratory sessions providing an introduction to the tools and concepts common to most areas of engineering; there are also group outings to engineering laboratories and industrial plants. Each student also concentrates on one particular topic, such as Optics and Lasers or Bioengineering, and meets one on one with a university professor to complete a project on that topic. The tuition cost for the Making of an Engineer course is only around $70, which includes field trips and supplies. Students can choose to reside on the University of Denver campus and pay an additional fee of about $800 for room, board, and social activities. Scholarships are available to needy students to cover residential and/or travel costs. Interested students must submit a completed application (with essay), official high school transcript (which must include at least one algebra course), standardized test results (PACT/ACT/PSAT/SAT), and letter of recommendation from a counselor or teacher. All materials should be submitted by the end of April to the Early Experience Program Coordinator, who can also provide application forms and further information about the Making of an Engineer course.

■**The Making of an Engineer**
Center for Educational Services

University of Denver
2135 East Wesley Avenue
Denver, CO 80208
Tel: 303-871-2663

MENTORING AND ENRICHMENT SEMINAR IN ENGINEERING TRAINING

College Course/Summer Study

The Cullen College of Engineering at the University of Houston offers the Mentoring and Enrichment Seminar in Engineering Training (MESET) for rising seniors during three weeks in June. Students planning a career in engineering who have demonstrated scientific and mathematical aptitude are eligible for this residential program. All participants attend short, formal courses in such subjects as computers, physics, problem solving, and engineering design. There are also a number of field trips to industrial facilities on the Gulf Coast and tours of Cullen College's own facilities. On evenings and weekends, students can enjoy social events including games and picnics and also attend lectures by guest speakers. There are no tuition or room and board fees for MESET, but participants must cover their own travel costs, weekend meals, and incidental expenses. A few transportation scholarships are available for cases of extreme hardship. To apply, students must submit an application as well as a high school transcript and letter of recommendation from their guidance counselor by a deadline that is usually in early April. For an application or more information about the program, contact the Director of MESET.

■ **Mentoring and Enrichment Seminar in Engineering Training**
PROMES
University of Houston
Houston, TX 77204-4790
Tel: 713-743-4222

MINORITIES AND WOMEN IN ENGINEERING, AND SUMMER YOUTH PROGRAMS

College Course/Summer Study

Michigan Technological University (MTU) offers three different opportunities for high school students to explore engineering in a college setting. The residential Minorities in Engineering Program is for minority or economically disadvantaged students who show talents in math and science; rising sophomores, juniors, and seniors are eligible. While living on campus for one week, participants engage in informational discussions and technical projects with MTU faculty and other professional engineers. The program is in mid-June and applications are due by April. The only cost for the program is a $50 registration fee.

The Women in Engineering Program is almost identical to the Minorities Program except, of course, that only female students are eligible. The registration fee is $100, payable upon acceptance.

Finally, Michigan Technological University offers the Summer Youth Program for all students between the ages of twelve and eighteen. Participants attend one of four week-long sessions held during the month of July, choosing either to commute or to live on campus. Students undertake an Exploration in one of many career fields—including engineering—through laboratory work, field trips, and discussions with MTU faculty and other professionals. The cost of the Summer Youth Program is $415 for the residential option, $235 for commuters. Applications are accepted up to one week before the Exploration begins.

■ **Minorities in Engineering, Women in Engineering, and Summer Youth Programs**
Michigan Technological University Youth Programs
1400 Townsend Drive
Houghton, MI 49931-1295
Tel: 906-487-2219

MINORITY ENGINEERING SUMMER RESEARCH PROGRAM
College Course/Summer Study

Vanderbilt University offers a limited number of rising seniors the opportunity to participate in its Minority Engineering Summer Research Program, which runs for five weeks beginning in early July. Originally designed for students entering Vanderbilt the following autumn, the program gives all of its participants a preview of college life while preparing them for a demanding engineering curriculum. Students concentrate on one of seven areas of engineering and are assigned to a summer school class that meets on weekday mornings. The afternoons are spent assisting a Vanderbilt professor on his or her latest engineering research project. Throughout the program, participants attend special lectures from business leaders and go on field trips to see the latest industrial technology firsthand. There is no cost for the Summer Research Program; Vanderbilt University provides room, board, transportation to and from Nashville, and a stipend to compensate students for being unable to work during the five-week period. Rising seniors who wish to apply must submit three letters of recommendation from math and science teachers, transcripts from grades nine and ten, and a completed application form. Applications are due by March 1, recommendations by April 1. For a copy of the form and for further information, contact the Assistant Dean for Minority Affairs at the Vanderbilt University School of Engineering.

■ **Minority Engineering Summer Research Program**
Vanderbilt University School of Engineering
Box 6006, Station B
Nashville, TN 37235
Tel: 615-322-2724

MINORITY INTRODUCTION TO ENGINEERING (MITE) AT AUBURN UNIVERSITY
College Course/Summer Study

Since 1978, the College of Engineering at Auburn University has sponsored the Minority Introduction to Engineering (MITE) for rising seniors from minority ethnic groups. Auburn's MITE program runs for three separate one-week sessions in the summertime, with about 25 students attending each session. Each student selected for MITE has an outstanding academic record (a GPA of 3.0 or better), the potential to score at least a 27 on the ACT, and a genuine interest in engineering, mathematics, or the sciences. During the program, participants learn about career opportunities in engineering and how their interests might suit those opportunities. By participating in classes on engineering concepts and meeting with present engineering students and faculty, participants begin to understand the demanding nature of a college major—and a career—in engineering. By living, eating, and socializing on the Auburn University campus, you can experience college life as a whole. Additional field trips, special projects, and hands-on experiences encourage students to find enjoyment in this career field and to consider meeting the challenges of engineering. MITE participants who go on to study engineering at Auburn are eligible for scholarships and internships sponsored by the Coca-Cola Company in cooperation with the university. For details on costs and application procedures, contact the Director of the Minority Introduction to Engineering Program.

■ **Minority Introduction to Engineering (MITE)**
211 Aerospace Engineering Building
Auburn University, AL 36849
Tel: 334-844-6820

MINORITY INTRODUCTION TO ENGINEERING (MITE) AT TUSKEGEE UNIVERSITY
College Course/Summer Study

The College of Engineering, Architecture and Physical Sciences at Tuskegee University invites minority students who are rising seniors to apply to the Minority Introduction to Engineering (MITE). MITE is a one-week summer program (check with Tuskegee University for more specific dates) that allows students to fully experience campus life, from living in the dormitories to studying with current staff and students. As a participant, you spend most of the week exploring engineering and other math and science careers and attending laboratory demonstrations by engineering faculty. You also go on field trips and hear lectures about college admissions, financial aid, and Army and Air Force ROTC options. There is no cost for this program except for transportation to and from Tuskegee University. For further details and application information, contact the Office of the Dean at the College of Engineering, Architecture and Physical Sciences. Minority students should also contact the Office of the Dean to enquire about Tuskegee University's other pre-college engineering programs, which vary from year to year.

Minority Introduction to Engineering (MITE)
Tuskegee University
College of Engineering, Architecture, and Physical Sciences
Tuskegee Institute, AL 36088
Tel: 205-727-8970 or 205-727-8430

NATIONAL ASSOCIATION OF PRECOLLEGE DIRECTORS

College Course/Summer Study, Field Experience

The National Association of Precollege Directors (NAPD) is a nonprofit group trying to increase the number of ethnically underrepresented students (Latino/Hispanic American, African-American, Native American) who pursue college degrees in engineering, mathematics, and technology. Its programs— collaborations between high schools, universities, and corporations—include in-school math instruction to ensure that students are on a college-entrance track; science and technology projects conducted with professors or engineers; students "shadowing" professionals throughout a day at work; and intensive summer programs on college campuses. To get a sense of the incredible variety of participants and projects, visit the Web site to view a roster of participating schools and click on some of the links. Or contact the Chairman of the NAPD, Robert Willis, at the address below.

National Association of Precollege Directors
The Johns Hopkins University
Applied Physics Laboratory
Johns Hopkins Road
Laurel, MD 20723-6099
Web: http://www.jhuapl.edu/NAPD/overview.htm

NATIONAL HIGH SCHOOL INSTITUTE—ENGINEERING SCIENCE DIVISION

College Course/Summer Study

The National High School Institute (NHSI), founded in 1931, is a college-based academic program for outstanding high school students. Its Engineering Science Division is based at Northwestern University, where rising seniors spend the month of July exploring the different aspects of the field of engineering. Students spend most of the day attending five mini-courses they choose from such topics as "Numerical Methods with Engineering Applications" and "Astronomy and Space Science." Participants also take time each day to work on their own research projects, which they create and pursue under the direction of a Northwestern faculty member; previous topics have included "Ternary Electronics" and "Women and the Culture of Science and Engineering." Throughout the program there are special lectures and field trips to a number of museums and research facilities in nearby Chicago. All participants in the NHSI Engineering Science Division live and study at Northwestern's Evanston campus, where there is ample opportunity for sports, arts, and leisure activities. The program fee is $2,775, which covers tuition, room, board, use of the university health service, field trips, and program-sponsored events and social activities. A number of scholarships and financial assistance options are available, each awarded on the basis of financial need and academic merit. Applicants should submit their financial aid request along with their application form, letter of recommendation, and a transcript including their GPA, class rank, and PSAT/SAT/ACT scores. The application deadline is in late April; those who meet the early admission deadline in late March will have a decision by late April. For further information, including a current listing of mini-course topics, contact the NHSI Engineering Science Division.

> ■ **National High School Institute—Engineering Science Division**
> Northwestern University
> 617 Noyes Street
> Evanston, IL 60208-4165
> Tel: 847-491-3026 or 800-662-NHSI
> Web: http://www.nwu.edu/summernu/special.html
> Email: nhsi@nwu.edu

NATIONAL SOCIETY OF BLACK ENGINEERS

Membership

The National Society of Black Engineers (NSBE) welcomes young people as members and as participants in its Pre-College Initiative (PCI) program. The PCI program links professional NSBE members with students in kindergarten through twelfth grade to encourage their interests in math and science. PCI student in grades seven through twelve are eligible to become NSBE Jr. members. Membership dues are only $5 per year, for which you can join or start a local chapter and participate in such things as "camping conferences" and college admissions and financial aid workshops. The National Society of Black Engineers urges you to call or email them if you want to make the NSBE part of your career preparation.

National Society of Black Engineers
1454 Duke Street
Alexandria, VA 22314
Tel: 703-549-2207, ext. 305
Web: http://nsbe.org/
Email: nsbejr@nsbehq.nsbe.org

OPERATION CATAPULT

College Course/Summer Study

Operation Catapult is a summer program operated by the Rose-Hulman Institute of Technology. Founded in 1874 under a different name, Rose-Hulman is today one of the country's best undergraduate colleges in science and engineering. It is a relatively small facility with a teaching (not research) faculty and is extremely selective in admitting students. Rose-Hulman's Operation Catapult is equally selective, usually extending invitations only to high school juniors who have finished in the highest percentiles on such standardized tests as the PSAT. Participants must also have completed three years of high school math and one of chemistry or physics.

There are two sessions each summer—generally the last three weeks of June and of July—with some ninety-five students participating in each session. Working in groups of two to four, participants try to solve a "real-world" problem in engineering or another scientific field. They are assisted by Rose-Hulman's faculty and upperclass students and make use of the institute's excellent facilities. Throughout Operation Catapult, students also go on field trips, witness demonstrations, and hear lectures. The program is a chance to live, work, and play on a college campus as well as to meet other students from around the country. Costs total about $1,400. Operation Catapult is obviously not for everyone interested in engineering, but if you meet the academic requirements, it's a unique and challenging opportunity that merits serious consideration. Eligible students should receive an invitation from Rose-Hulman Institute of Technology; if you believe you are qualified but have not been invited, ask your guidance counselor or a teacher for direction.

■**Operation Catapult**
Rose-Hulman Institute of Technology
5500 Wabash Avenue
Terre Haute, IN 47803

PCS SUMMER CAMPS

Camp

PCS Education Systems runs an assortment of summer camps at various locations in the western part of the country. Most of their senior camps are for young people over the age of ten; some cut off eligibility at age sixteen, others accept campers up to age eighteen. Senior camps are offered throughout July and August in Corte Madera, California; Spokane, Washington; Redmond, Washington; San Rafael, California; and Boise, Idaho. These are not residential programs, but meet for three or four hours every afternoon for one week. Prospective engineers have several options at most of these camp sites. You can explore math, physics, and other engineering concepts through LEGO construction at such camps as the LEGO Mechanical Engineering Challenge and LEGO (Un)Civil Engineering. Other possibilities include RoboCamp (study and experimentation with robotics), Adventures in Electricity (the basics of electronics and magnetics), and Rocketeers Camp (a comprehensive introduction to rocketry and even parachute design). The cost of each senior camp is, on average, about $150. Most camps have no formal application deadline, but be sure to apply early, as they fill up quickly. Contact the PCS general office at the address given here for detailed information on the summer camp in your area.

■**PCS Summer Camps**
PCS Education Systems
1444 West Bannock, PO Box 3298
Boise, ID 83703
Tel: 208-343-3110
Web: http://www.pcsedu.com

PRE-COLLEGE ENGINEERING PROGRAM (PREP)

College Course/Summer Study

The College of Engineering at the Georgia Institute of Technology offers rising juniors and seniors the chance to participate in its PRe-college Engineering Program (PREP). This residential program is run in two one-week sessions at the end of June. Thirty males and thirty females take part in tours of the institute's engineering schools, overviews of its curriculum, and panel discussions with its students, as well as visits to local and out-of-state industries, film and slide presentations, and a group engineering project. You close out the week with workshops and a career fair featuring major corporations. Participants live and eat on campus, although everyone dines out when on a field trip. All meals plus accommodation, tuition, and excursions are covered in the program fee of $600; students need only pay for travel to and from Atlanta and incidental expenses. To apply, you must complete a form and submit it along with a high school transcript, letter of recommendation, and a personal statement by the end of March. The main factors influencing admission are GPA, curriculum, and PSAT/SAT/ACT scores. Students who have already participated in a program similar to PREP are not eligible. For details and an application form, contact the Assistant Dean of the College of Engineering. The Georgia Institute of Technology offers a range of engineering programs for rising sixth-graders through rising seniors. The Assistant Dean of the College of Engineering can provide further information.

■ **Pre-College Engineering Program (PREP)**
College of Engineering
Office of Minority and Special Programs
Georgia Institute of Technology
Atlanta, GA 30332-0361
Tel: 404-894-3354

PRE-FRESHMAN ENGINEERING PROGRAM (PREP)
College Course/Summer Study

The Rochester Institute of Technology (RIT), in cooperation with the University of Rochester, offers the PRe-freshman Engineering Program (PREP) for current eighth- and ninth-graders living in the Rochester City School District. PREP was established in 1988 and has since sent nearly forty graduates on to major universities, with almost half of them studying science and engineering. Twenty-five minority students are selected to participate in this two-part program each year based on academic excellence and strong math and science skills. The first part is a two-week commuting program held in July at the RIT campus. Students work with the faculty there on individual and group projects in such disciplines as chemical, civil, electrical, and mechanical engineering. The July session also includes field trips, usually to the Mobil Chemical Company and the Eastman Kodak Company. The second part of PREP consists of five seminars held on Saturdays throughout the school year. Seminars include additional field trips and advice on preparing for college. From completion of the program until high school graduation, PREP students are eligible for scholarships for summer classes at RIT and the University of Rochester and to serve as teaching assistants in subsequent summer PREP sessions. Students selected to participate in PREP pay no fees for the program; even bus transportation to and from the RIT campus in the summer is provided. Applications are due by the end of March; selected students will be called back for interviews in April. For more information and an application, contact the RIT Department of Mechanical Engineering.

▓ **Pre-Freshman Engineering Program (PREP)**
Rochester Institute of Technology
Department of Mechanical Engineering
76 Lomb Memorial Drive
Rochester, NY 14623-5604
Tel: 716-475-2162

PREFACE AND MITE AT PURDUE UNIVERSITY

College Course/Summer Study

Purdue University's Department of Freshman Engineering offers two programs for minority high school students in conjunction with the University's Minority Engineering Program. The first program, Preface (Pre-Freshman and Cooperative Education), is for rising juniors and seniors who show promise in math and science. Preface runs for one week in July, during which participants explore possible engineering careers, meet professional engineers and engineering students, and work on study skills and career and life planning. There are also tours, hands-on activities, and a group project. Participants live on the Purdue campus. Tuition, room, and board costs are fully funded by Purdue's industrial partners, so Preface students are responsible only for the costs of transportation, incidentals, and a $150 administration fee. Applications are due roughly one month before the program begins, but should be sent in earlier because of the popularity of this program.

Purdue University's second program is MITE (Minority Introduction to Engineering), which runs for two weeks in mid-July. Rising seniors with a strong academic background take part in computer sessions, laboratory experiences, engineering design projects, and lectures by faculty members and engineering professionals. Participants live on campus and have access to a wide range of recreational events and facilities. Like Preface, MITE is fully funded, so students pay just for travel to and from Purdue, incidentals, and an administration fee of $175. There is no formal admissions deadline, but students should apply as soon as the forms become available in May because eligible applicants are accepted on a first come, first served basis. For more information about both programs and details about application procedures, contact the Department of Freshman Engineering.

Purdue University also offers Summer Engineering Workshops for seventh- and eighth-graders in June of each year. They last for about one week and provide an initial exposure to the field of engineering. Contact the Department of Freshman Engineering for details.

Preface and MITE Programs
Purdue University
Department of Freshman Engineering, 1286 ENAD
West Lafayette, IN 47907-1286
Tel: 765-494-9713

PRIME, INC.

College Course/Summer Study

PRIME is a nonprofit organization dedicated to "creating opportunities for minorities in mathematics- and science-based professions." It serves schools in the Delaware Valley, namely those in the Philadelphia, Camden, Norristown, and William Penn school districts, the Archdiocese of Philadelphia, and selected private schools. Minority high school students can participate in exciting math and science activities as well as career exploration through the PRIME Academic Year Program (running from autumn to spring) and the PRIME Universities Program (running throughout the summer at regional universities such as Penn State and Drexel). PRIME also sponsors the Saturday Tutorial and Enrichment Program and the Providing Activities for Careers in Technology course for high school students who need supplementary instruction. PRIME was initially founded in 1973 to encourage minority students to enter the field of engineering, so you can be confident of the organization's commitment to help you reach your career goals. There is no cost to you for any PRIME program, so if you are interested in joining one and you attend one of the schools served by PRIME, speak to your science teacher or guidance counselor about getting involved. Many of PRIME's programs are also extended to middle school students, some even to fifth-graders.

PRIME, Inc.
The Wellington
135 South 19th Street, Suite 250
Philadelphia, PA 19103-4907
Tel: 215-561-6800

SCIENCE OLYMPIAD

Competition

The Science Olympiad is a national competition based in schools. School teams feed into regional and state tournaments, and the winners at the state level go on to the national competition. Some schools have many teams, all of which compete in their state Science Olympiad. Only one team per school, however, is allowed to represent its state at the national contest, and each state gets a slot. There are four divisions of Science Olympiad: Division A1 and A2 for younger students, Division B for grades six through nine, and Division C for grades nine through twelve. There is no national competition for Division A.

A school team membership fee must be submitted with a completed membership form thirty days before your regional or state tournament. The fee entitles your school to a copy of the Science Olympiad Coaches and Rules Manual plus the eligibility to have up to fifteen students at the first level of your state or regional contest. Fees vary from state to state. The fee to enter the national competition is $60, but some states may add an additional cost. The National Science Olympiad is held in a different site every year, and your school team is fully responsible for transportation, lodging, and food.

Specific rules have been developed for each event and must be read carefully. There are twenty-two different events in each division. You and your teammates can choose the events you want to enter and prepare yourselves accordingly. Winners receive medals, trophies, and some scholarships.

For a list of all Science Olympiad state directors and a membership form, go to the Science Olympiad Web site. You can also write or call the national office for information.

▉ Science Olympiad
National Office
5955 Little Pine Lane
Rochester, MI 48306
Tel: 248-651-4013
Web: http://www.geocities.com/CapeCanaveral/Lab/3812/what.html

SECONDARY STUDENT TRAINING PROGRAM (SSTP) RESEARCH PARTICIPATION

College Course/Summer Study

The University of Iowa invites those who have completed grade ten, eleven, or twelve to apply to its Secondary Student Training Program (SSTP), now in its thirty-ninth year. The program allows students to explore a particular area of science, such as engineering, while experiencing the career field of scientific research. Participants work with university faculty in one of the many laboratories on campus, studying and conducting research projects for approximately forty hours per week. At the end of the program, which runs from late June to early August, you present your project to a formal gathering of faculty, staff, and fellow SSTP participants. Throughout the program you also take part in various seminars on career choices and the scientific profession, and a variety of recreational activities designed especially for SSTP participants. Students live in the University of Iowa dormitories and use many of the campus' other facilities. The admissions process is highly competitive and is based on an essay, transcript, and recommendations. Those who complete the program receive college credit from the University of Iowa. Applications are due by mid-March, and applicants will be notified of the decisions by mid-May. Tuition fees, room, and board total $1,650; spending money and transportation to and from the university are not included. For an application form and to discuss possible research projects, contact the Secondary Student Training Program.

■ **Secondary Student Training Program (SSTP) Research Participation**
University of Iowa, Iowa SSTP
323 Chemistry Building D
Iowa City, IA 52242-1294
Tel: 319-335-0040

SOCIETY FOR MINING, METALLURGY, AND EXPLORATION

Competition

The Society for Mining, Metallurgy, and Exploration (SME) holds an annual photo contest seeking photographs that promote the minerals industry. Entries are accepted in black-and-white and in color in two divisions: open and student (all those under nineteen years of age). Subject matter must be related to mining and can include such themes as mine restoration, environmental practices, smelting and refining, and surface or open-pit landscapes. Entrants can submit up to four photographs: two in black-and-white and two in color. Winning entries are awarded cash prizes for first ($150), second ($100), and third ($75) places; honorable mention ($50) may be given at the judges' discretion. The SME may exhibit or otherwise use the winning photographs in its work, but the photographers retain all of their rights to the photos. For more information on required format, submission procedures, and an entry form, contact the Government, Education, and Mining (GEM) Committee of the SME.

■ **Society for Mining, Metallurgy, and Exploration**
8307 Shaffer Parkway
Littleton, CO 80127

SOCIETY OF WOMEN ENGINEERS HIGH SCHOOL CONFERENCE

Conference

The Society of Women Engineers (SWE) at Texas A&M University welcomes high school students at all grade levels to its annual High School Conference.

Each February, conference participants spend a weekend on the Texas A&M campus learning about the many engineering disciplines and seeing what the university has to offer. On Saturday, students tour the campus, attend typical classes, participate in a design competition, and attend a seminar on each engineering discipline. At the end of the day, everyone attends a banquet dinner at which certain graduating seniors who will be attending Texas A&M in the fall receive scholarships. On Sunday, students tour the departments of engineering that most interest them and learn more about those disciplines. For more information about the conference, including specific dates and times, contact SWE's High School Conference Chairwomen.

The Society of Women Engineers at Texas A&M University is in the process of planning a summer camp for young women who are rising juniors and seniors in high school. Contact SWE for the latest information.

Society of Women Engineers High School Conference
College of Engineering
Texas A&M University
College Station, TX 77843-3127
Tel: 409-862-2314

STUDENT INTRODUCTION TO ENGINEERING AT NORTH CAROLINA STATE UNIVERSITY

College Course/Summer Study

The Student Introduction to Engineering (SITE) is a summer program sponsored by the College of Engineering at North Carolina State University. Open to rising juniors and seniors from all backgrounds, SITE offers high school students a realistic look at the professional lives of engineers and the preparation needed to pursue such a career. As a participant, you spend one week living on the NCSU campus and participating in discussions and demonstrations in its lecture halls and laboratories. You also meet with engineering students and practicing engineers to learn what you can expect in the years to come. North Carolina State University provides information on college admissions and financial aid to all the students in the SITE program. The cost of participation is about $300, which includes room, board, tuition, gymnasium access, and insurance coverage. Some financial assistance is available. The same material is covered in two different SITE sessions, both of which run in June. To apply, select the session you want and submit the application form by mid-April. Applicants are judged on the basis of scholastic performance, interest in math and science, teacher recommendations, and standardized test scores. For more information and a copy of the application form, contact the SITE Coordinator.

▪Student Introduction to Engineering (SITE) at North Carolina State University
NCSU College of Engineering
PO Box 7904
Raleigh, NC 27695-7904
Tel: 919-515-9669

SUMMER COLLEGE FOR HIGH SCHOOL STUDENTS AT SYRACUSE UNIVERSITY
College Course/Summer Study

The Syracuse University Summer College for High School Students features an Engineering and Computer Science Program for those who have just completed their sophomore, junior, or senior year. The Summer College runs for six weeks from late June to early August and offers a residential option so participants can experience campus life while still in high school. The Engineering and Computer Science Program has several aims: to introduce you to the many specialties within the engineering profession; to help you match your aptitudes with possible careers; and to prepare you for college, both academically and socially. Participants take two courses; everyone takes the Survey of Engineering Problems course and selects a liberal arts and sciences course to complement it. The faculty recommends the selection of a mathematics, computer, or physics course. All students take part in a number of engineering-related field trips and group design projects, the results of which are presented at the end of the course. Syracuse University awards college credit for completion of the two courses. Admission is competitive and is based on recommendations, test scores, and transcripts. The total cost of the residential program is about $3,500; the commuter option costs about $2,500. Some scholarships are available. The application deadline is in early June, or early May for those seeking financial aid. For further information, contact the Summer College.

▪Summer College for High School Students
Syracuse University
Summer College Programs, 309 Lyman Hall
Syracuse, NY 13244-1270
Tel: 315-443-5297
Web: http://www.syr.edu/WWW-Syr/Academic Life/HighSchool/index.htm

SUMMER ENGINEERING EXPLORATION & SHADOW DAY
Camp, Field Experience

The Society of Women Engineers at the University of Michigan hosts a Summer Engineering Exploration (SEE) camp and a Shadow Day experience. SEE is a one-week residential camp held in late July or early August for all interested high school girls. While living on campus and exploring undergraduate life at the University of Michigan, participants learn about many different areas of engineering. Learning activities include lectures, lab tours, a design project, and a corporate tour. The cost is about $225, which mainly goes toward room and board. For more information and an application form, contact the Outreach Director. During the school year, high school girls can take part in a Shadow Day—there is usually one in October and one in March. Participants "shadow" a female engineering student at the University of Michigan for the whole day. The itinerary usually includes breakfast, a campus tour, a panel discussion of academic and extracurricular activities, lunch, and a typical day of classes. The cost is only $5, which pays for lunch. If you are interested in a Shadow Day, contact the High School Relations Chair.

> ■ **Summer Engineering Exploration & Shadow Day**
> Society of Women Engineers
> University of Michigan, 1226 EECS
> Ann Arbor, MI 48109
> Tel: 313-763-5027

SUMMER HIGH SCHOOL ENGINEERING INSTITUTE

College Course/Summer Study

Since 1963, the College of Engineering at Michigan State University (MSU) has offered the High School Engineering Institute to rising juniors and seniors each summer. Taking place in July, the program gives students with strong interests in math and science the chance to explore engineering as a career. Participants live on campus and participate in hands-on activities, discussions, and laboratories with MSU faculty members and graduate assistants. During the course of the institute, you are exposed to all ten engineering disciplines offered at the university, including biosystems, civil, electrical, and mechanical. There are also opportunities to experience the social life on campus, enjoy the recreational facilities, and learn more about the entire college admissions process. Admittance to the program is subject to considerations of academic performance, class rank, and the recommendation of a teacher or counselor. Applications must be submitted by mid-May along with a $50 deposit (to be refunded if you are not accepted into the institute). Costs of the program total around $350, from which the deposit is deducted. A limited amount of financial aid is available to those students otherwise unable to attend. Contact the High School Engineering Institute for the latest information and an application.

> ■ **Summer High School Engineering Institute**
> Office of Undergraduate Studies, College of Engineering
> Michigan State University
> East Lansing, MI 48824-1226
> Tel: 517-355-6616

SUMMER INSTITUTE FOR CREATIVE ENGINEERING AND INVENTIVENESS (SICEI)
College Course/Summer Study

The University of Iowa offers a two-week, residential Summer Institute for Creative Engineering and Inventiveness (SICEI). The program, held in July, is sponsored by the Belin-Blank International Center for Gifted Education and Talent Development, part of the university's College of Education. The SICEI is open to students entering the ninth, tenth, or eleventh grade in the autumn. Applicants to the SICEI must also be members of NRPHSS or BESTS (these are talent searches) or have been a Blank Scholar or Governor's Scholar. Those who are selected for the program live on the University of Iowa campus and work in teams to explore the field of engineering and the engineering design process itself. Participants work on problem solving, research methods, establishing objectives and criteria, and preparing reports and presentations. The program cost of about $1,000 covers tuition, room, board, recreation activities, and other fees. For further details and for information about the application process, contact the Program Administrator of the Summer Institute for Creative Engineering and Inventiveness.

> ■ **Summer Institute for Creative Engineering and Inventiveness (SICEI)**
> Belin-Blank Center, College of Education
> 210 Lindquist Center
> Iowa City, IA 52242-1529
> Tel: 800-336-6463 or 319-335-6148

SUMMER INSTITUTE IN SCIENCE AND ENGINEERING
College Course/Summer Study

The Summer Institute in Science and Engineering for High School Juniors is held on the Alfred University campus for one week at the end of June. By "juniors," the university means both rising juniors and those who have just completed their junior year (i.e., rising seniors). Participants live in campus dormitories and participate in seven hands-on labs they have selected from a choice of about twenty-five; past lab topics have included ceramic manufacturing, genetic engineering, and cryptography. Students must apply for the Summer Institute by the beginning of May, submitting an application form, transcript, $25 deposit, and two letters of recommendation from high school teachers. The total cost for tuition and room and board is around $325, and some financial aid is available. Participants who submit a paper on a given topic before the first day of the institute will be in competition for a four-year scholarship to Alfred University, the winner to be decided based on the quality of the writing. Contact the Director of the Summer Institute for more information.

■ **Summer Institute in Science and Engineering**
New York State College of Ceramics at Alfred University
2 Pine Street—BMH 215
Alfred, NY 14802-1296
Tel: 607-871-2425
Fax: 607-871-2392

SUMMER SCHOLARS IN COMPUTER SCIENCE AND ENGINEERING PROGRAM

College Course/Summer Study

The Summer Scholars in Computer Science and Engineering Program is sponsored by the University of Maryland and Howard University for rising seniors
from underrepresented ethnic groups (African-Americans, Latino/Hispanic
Americans, and Native Americans). Participants in the five-week program
spend July exploring career opportunities and taking college-level courses at
both universities. The first course is "Introduction to Engineering Design," in
which students study the design process and problem solving, and complete a
group design project. The other course is "Introduction to Information
Technology", a hands-on class that focuses on concepts of computing and
related terminology. College credit is awarded to students who satisfactorily
complete both courses. Participants live in the residence halls at Howard
University and take their meals on campus. Tuition, room, books, and lunches
Monday through Friday are provided at no cost to participants; the only cost is
for other meals and for transportation to and from Howard University. To
apply, you must submit an application form, an essay, two letters of recommendations, and a current transcript by mid-April. Admissions decisions are
based primarily on the recommendations, GPA, standardized test scores
(PSAT/SAT), and overall academic ability. For further details and an application
form, contact the Assistant Director.

■ **Summer Scholars in Computer Science and Engineering Program**
Center for Minorities in Science and Engineering
Room 1134, Engineering Classroom Building
University of Maryland
College Park, MD 20742
Tel: 301-405-3878
Web: http://www.engr.umd.edu/organizations/cmse/

SUMMER STUDY IN ENGINEERING FOR WOMEN HIGH SCHOOL STUDENTS

College Course/Summer Study

The University of Maryland at College Park (UMCP) runs the Summer Study in
Engineering for Women High School Students for six weeks in July and August.
Rising seniors who are considering careers in engineering learn more about the
various disciplines within the field while earning credit through this program,

which has existed since 1975. Participants take two regular college classes, the first of which is "Introduction to Engineering Design." This course covers the fundamentals of engineering and reinforces them through hands-on group design projects. The other class, "The World of Engineering," introduces students to the nine engineering disciplines offered at UMCP—including aerospace, fire protection, and nuclear—via visits to each department and field trips to government labs and industrial firms. Outside of class, you take part in other engineering-related tours, meetings, and workshops as well as recreational activities. Participants live on the UMCP campus in apartment-style dorms and are free to participate in campus events as long as they do not conflict with the Summer Study in Engineering. All costs are covered by the university and other organizations, except for meals, transportation to and from UMCP, and a $100 registration fee. To be considered for the program, the university must receive a completed application form, a letter of recommendation, and a high school transcript by early May. For more information, contact the program directly.

■ **Summer Study in Engineering for Women High School Students**
Women in Engineering Program
University of Maryland
1131 Engineering Classroom Building
College Park, MD 20742
Tel: 301-405-0315

UNIVERSITY OF WISCONSIN-STOUT T.C.E.A. HIGH MILEAGE VEHICLE

Competition

This engineering competition to research, develop, and design a single-person, fuel-efficient vehicle out of a Briggs & Stratton engine takes place every spring. Facilitated by University of Wisconsin-Stout students of the Technology Collegiate Education Association (T.C.E.A.), the contest has been running since 1991 and draws from high schools all over Wisconsin. However, any school is eligible to sponsor a team, which usually consists of about five people. To enter, all you have to do is transport your car to the competition site. The competition is held over a Friday and Saturday. Cars roll in on Friday and go through qualifying tests for safety, durability, and weight and stoppage requirements. On Saturday, the competition begins at a nearby industrial park. Winners receive trophies in four categories: stock class, modified class, team work, and design. The sponsoring school usually helps teams with the cost of transportation and lodging, and many teams solicit the help of businesses and corporations in their town. The entry fee is about $30. To receive information about the competition, contact the Communications Education and Training Department at the University of Wisconsin-Stout.

■ **University of Wisconsin-Stout T.C.E.A. High Mileage Vehicle**
Menomonie, WI 54751
Tel: 715-232-1206
Email: weltyk@uwstout.edu

VISIT IN ENGINEERING WEEK (VIEW) SUMMER PROGRAM
College Course/Summer Study

The Pennsylvania State University (Penn State) invites rising juniors and seniors to apply to its Visit in Engineering Week (VIEW) residential summer program. Participants must be academically talented, motivated, genuinely interested in engineering, and members of an underrepresented minority group: Native American, African-American, Latino/Hispanic American. There are three week-long VIEW sessions each summer, usually two in July and one in August. Each session provides experiences in design, modeling and implementation, communications, group dynamics, and project management. Participants explore many different areas within the field of engineering and also sample college life as an engineering student. All fees, including tuition, and room and board, are underwritten by corporate sponsors, so VIEW is free to all students. You must submit an application form, transcript, and essay by the end of April to the Program Director, who is also available to answer your questions.

■ **Visit in Engineering Week (VIEW) Summer Program**
The Pennsylvania State University
241 Hammond Building
University Park, PA 16802
Tel: 800-848-9223

WESTINGHOUSE SCIENCE TALENT SEARCH
Competition

Since 1942, the Westinghouse Electric Corporation has held a nationwide competition for talented high school seniors who plan to pursue careers in science, engineering, math, or medicine. Those who win find themselves in illustrious company: former winners have gone on to win Nobel Prizes, National Medals of Science, MacArthur Foundation fellowships, and memberships in the National Academy of Engineering. Here's how it works. In 1996, fifteen hundred students nationwide entered projects. These were judged by ten scientists who chose three hundred as semifinalists to receive recommendations to colleges and universities. Next, forty finalists were selected and given all-expense-paid trips to Washington, DC, where they met with world-class scientists as well as the president, and displayed their exhibits at the National Academy of Sciences. Finally, on the basis of interviews, ten top scholarship winners were chosen and awarded funds ranging from $10,000 to $40,000. The other thirty finalists all received $1,000 scholarships. If you'd like to enter next year's competition, contact the Talent Search for more information.

Westinghouse Science Talent Search
PO Box 4400
Flushing, NY 11356
Tel: 800-292-4452

WOMEN IN THE SCIENCES AND ENGINEERING (WISE) WEEK
College Course/Summer Study

The Pennsylvania State University (Penn State) offers a Women in the Sciences and Engineering (WISE) Week program in mid-June for female rising seniors. Participants are academically talented with strong math and science skills, headed for college, and considering their career path. Students apply to one WISE option, either Sciences or Engineering. Competition is considerable as only thirty-six young women are accepted into each option. Once in the Engineering option, participants take part in eight engineering and two science workshops while completing a week-long engineering design project. During WISE Week you also meet female role models in academic and industrial engineering and learn about educational opportunities at Penn State. Accommodation is in a campus residence hall with collegiate women as your supervisors. The cost of the program is about $250, which covers everything except transportation to and from Penn State's University Park Campus; a limited number of need-based scholarships is available. A completed application form, one letter of recommendation, an essay, and a current high school transcript must be submitted by the beginning of April. Members of minority groups and students with physical disabilities are strongly encouraged to apply. For further information about WISE Week and the application process, contact the program.

■ **Women in the Sciences and Engineering (WISE) Week**
The Pennsylvania State University
510 Thomas Building
University Park, PA 16802-2113
Tel: 814-865-3342
Fax: 814-863-0085
Email: WISE@psu.edu

WORLDWIDE YOUTH IN SCIENCE AND ENGINEERING (WYSE)

College Course/Summer Study

The College of Engineering at the University of Illinois, Champaign-Urbana, sponsors a Worldwide Youth in Science and Engineering (WYSE) program twice each summer. Students who have completed their sophomore, junior, or senior year are eligible for the program, which is subtitled: "Exploring Your Options . . . Tomorrow's Careers in Science and Engineering." The two-week program features visits to each department within the College of Engineering, where faculty, graduates, and undergraduates conduct presentations, discussions, and hands-on activities concerning their particular area. Participants can make personal appointments with faculty members in engineering and nonengineering fields alike. A number of special seminars are included in the WYSE program, including ones on study skills, choosing the right college, and accessing the Internet. Students enjoy many special activities and events planned by WYSE and have access to the athletic facilities on the Urbana-Champaign campus. You also live and eat on campus. Participants are selected according to the following factors: GPA, class ranking, curriculum, PSAT/SAT/ACT scores, letters of reference, and a personal essay. Applications are due approximately one month before the session begins, but when all other deciding factors are equal, those applications received earliest take precedence. The all-inclusive fee for the program is $450. Some financial aid is available; African-American and Latino/Hispanic American students may apply for Minority Introduction to Engineering (MITE) scholarships, which cover the entire cost of the program. For more information about the WYSE program and an application, contact the Program Director.

■ **Worldwide Youth in Science and Engineering (WYSE)**
University of Illinois, Urbana-Champaign
1308 West Green Street, Room 207
Urbana, IL 61801
Tel: 217-244-4974 or 800-843-5410
Fax: 217-244-2488

Do It Yourself

Twenty-nine years ago, on July 20, 1969, a human being walked on the pocked, chalky-white surface of the moon for the very first time. As he dropped to the surface from the last rung of the lunar module, Neil Armstrong's historic words were immediately transmitted back to earth: "That's one small step for a man, one giant leap for mankind." Millions of people sat motionless before their radios and televisions, trying to wrap their minds around the enormity of what he had just uttered and, more importantly, from where his voice came.

In the years since Armstrong, Buzz Aldrin, and Michael Collins made that incredible journey to the moon and back, much has changed about our world, and some of those very changes are a direct result of the engineering genius that, through years of research, design, and development, sent three Americans to the moon. Today, for instance, government and private communications satellites orbit the earth, transmitting live broadcasts from all around the world; food products are sealed into special containers that don't require cooling—helping people who live in hot, isolated regions with no electricity; special, lightweight, synthetic materials protect mountain climbers and Indy 500 auto racers alike from exposure to heat and cold; and computers which used to be housed in several large rooms now fit inside a business executive's pocket (the chips which power them fit on the head of a pin). Even now, scientists are working to perfect a way of "farming" on space stations the antidotes to deadly viruses which are too expensive to produce on Earth.

The fantastic discoveries and developments which resulted from space exploration are, however, only one example of how the work of engineers affects our lives. Even something as seemingly simple as hopping on a bike for an afternoon ride demonstrates the degree to which engineering is involved with nearly every instant and every aspect of our lives. From the gears on your bicycle, to the reflective, plastic swoosh on your tennis shoes, to the dense styrofoam padding the inside of your bike helmet, to the LCD display on your watch—these were once the newfangled gadgets of the future, a brainstorm in the mind of an engineer somewhere. Engineers turn problems into solutions, setbacks into experiments. Remember, hundreds of years ago a man named Galileo was ostracized for believing the earth was round and talking about travel to the stars . . . it does take a certain amount of faith and imagination to be a renaissance man, as it does to be an engineer. But if you see opportunities where others see only obstacles, perhaps you're cut out to be an engineer, too.

Are you naturally curious, always asking questions? When something at home breaks, do you take it to a mechanic or repair technician, or do you pull out your tool box and try to fix it yourself? Do you enjoy taking things apart even if they're not broken just so you can see what the insides look like? Just to see if you can put it back together again? Do you actually look forward to solving problems, and feel exhilarated when you solve them?

John Stasey's father is a mechanical engineer, and already at fourteen, John knows he wants to follow in his dad's footsteps. John has made a hobby out of taking things apart and putting them back together again. He scours garage sales, looks in second-hand stores, even scouts out the trash, searching for old, broken mechanical objects—toasters, wall clocks, light fixtures, and lawn mowers—and once he finds them, he tries to repair them. This is often much easier than it might seem; in a disposable society people are so used to tossing something out once it stops working, rather than trying to fix it, that often, the device or machine needs little more than a little oil to lubricate the working parts. After taking them apart, John gives each item a complete overhaul, cleaning all of the individual pieces, oiling them as necessary. With certain items, John waits for his father's help, but he learns quickly, and rarely has to have his father show him how to make a repair twice. More and more, John works alone, saving the stubborn problems for when his father has time to help him. Once the items are repaired John saves them for the family's annual summer garage sale. At the last sale John made $230.

John's financial success quickly prompted a couple of his friends to join him in his hobby. Together, they work in John's basement after school and on the weekends, figuring that as a group they can collect more salvageable items, fix them faster, learn from each other's projects, and make a bigger profit at this summer's garage sale. Occasionally they work on items that they just simply can't fix, and often they don't realize this until they've spent a couple of days working on the problem. Although John admits this can be very frustrat ing, he emphasizes that it's definitely not a waste of time. "We learn as much or more from the things we can't fix as we learn from the things we do fix," he says.

At first glance, it might seem like John is just having a good time monkeying around with old, broken-down machines, and making a little spending money. Look a bit more closely and you'll realize that John is well on his way to becoming an engineer, if not a prosperous businessman. He has a working knowledge of machines and mechanics, is developing problem-solving and troubleshooting skills, and, now that he's working with a group of friends, he is developing the teamwork and communication skills crucial for success as an engineer. His hobby is helping him attain his dream of one day becoming a mechanical engineer. What are you doing to realize your dreams? What can you do now?

WHAT CAN YOU DO?

More than you may realize. There are many ways young people can get involved in the field of engineering. As you can see from the number of chapters in this book, engineering is not just a single career, although each and every branch of engineering is based on the same fundamental understanding of science and math. No matter which branch of engineering they want to specialize in, most engineers start by acquiring solid math and science skills, as well as some sort of experience that allows them to apply their math and science knowledge in a practical, hands-on manner. Depending on what type of engineer you want to be, your opportunities for experience and exploration will vary. If you look in the right places and are creative, you will see that it isn't difficult to locate many fun and helpful opportunities to learn engineering and science skills or, like John did, even create your own opportunities.

MATH

The first thing you can do is pay close attention in math class. All engineering fields use math. Lots of it. Math is one of the twin pillars of engineering, so if this isn't your favorite subject and you think you may want to be an engineer, then it's time to go in for extra help after class. No matter how good you are at fixing things, or understanding complex problems, if your math skills are poor, you can't be an engineer. Why? The scientific principles which govern engineering rely on the accurate execution of, you guessed it, mathematical equations and formulas. You can't do one without the other. Don't worry, though, with a little hard work and determination you can certainly conquer math. The more you do it, the easier it gets. Better yet, try and find ways to use your math skills, polished or not. Using practical, real examples to which you can apply the math is one of the best ways to make any subject come alive.

If you're already a whiz at math (maybe you even feel a tingle in your spine when you solve that crazy problem that has everyone else stumped), then you're well on your way. Keep it up. But don't grow complacent and let your spark fizzle; keep pushing yourself, keep giving your math-savvy mind new challenges.

MATH, SCIENCE, & ENGINEERING CLUBS

Many junior high and high schools have math and science clubs that meet after school. These are great places to advance your math and science skills beyond the classroom. And how about fueling a little healthy competition? Some clubs even participate in competitions with other local high schools. Talk to the math and science teachers at your school about getting involved. If your school doesn't already have a science or engineering club, start one yourself. Enlist the help of one of your math or science teachers; most likely, they'll be overjoyed to find a group of kids who are willing and interested in exploring math and science in more depth. They will probably be happy to assist, or at least steer you in the right direction. Like other school clubs or teams, a math, science, or engineering club needs the leadership of a teacher, especially if you're thinking of conducting experiments using school equipment and facilities. Just stop and think about that for a moment.

Imagine the super-cool experiments you and your fellow club members can conduct. You're in charge, so whatever you're interested in exploring, studying, proving—you can do. And because of the wide scope of engineering, there should be no shortage of topics your club can study and perform experiments on, topics like physics, electricity, chemistry, mechanics. Working with others on projects like these helps you develop teamwork and problem-solving skills.

Among the projects which math, science, and engineering clubs have explored are alternative and experimental means of transportation (flying cars), fuel (solar power), and elaborate Mousetrap-like games which demonstrate different fundamentals of science.

Randy Clifton found a project his high school engineering club could solve that would gain him the respect of his principal, the janitors, the basketball team, and the whole student body. It also impressed the school superintendent so much that Randy's club was granted extra funds to buy some equipment they would need to perform more experiments. Here's how it happened.

The high school had a relatively old electronic scoreboard in their gym for basketball games. The scoreboard had analog numbers that were controlled electronically. Over the summer it broke down and when one of the janitors went to take a look at it, he only made matters worse. A repair man was finally brought in to fix it, but the repair cost was too high at the time for the school's budget. The repair technician recommended that the school buy a new scoreboard. But that, too, was out of the question.

When school started that fall, just a few months before basketball season, Randy, along with his science teacher, the club's supervisor, persuaded the principal to let them have a look at the scoreboard. "It was a mess," Randy remembers. "It hadn't been properly maintained over the years so there was all kinds of crud and rust inside." Not to mention the additional damage done by the janitor.

Every day after school for a month the club's seven members went to work on the scoreboard. They pinpointed and identified the problems. They cleaned it and removed worn or damaged parts, replacing them with parts they made themselves in the shop room, or bought from the hardware store with money the principal donated. In the end repair costs were only fifty-seven dollars—five times less than the repair technician estimated it would cost. The scoreboard went back up just days before the first game of the season. The crowd rewarded Randy and his engineering club with a big round of grateful applause.

MUSEUMS

Many cities have science museums where young people and adults can participate in interactive exhibits that teach principles of science and engineering. Talk to one of your teachers about a field trip to the museum—maybe volunteer to coordinate a field trip with a particular lesson. Of course, you can always go on your own or with some friends on a weekend or day off. In these museums you can find fun and exciting exhibits that demonstrate principles of electricity, chemistry, and the environment; show the inner mechanics of complex machines; explore the intricacies of space travel; as well as many other scientific and technical topics.

Museums often provide opportunities for the curious young scientist and engineer beyond the museum's in-house exhibits. Some sponsor science clubs where you can have a much more active role in designing and running experiments. Some of the larger museums like the Museum of Science and Industry in Chicago have teen volunteer programs in which qualified teens work in facilitating museum-organized activities for younger children. This is a great way to get involved in the museum, to really learn a subject, and to meet people in the industry. Which brings up an interesting point: Not many people stop and think about where they want to work; in addition to jobs in the major industries, museums also employ engineers and scientists to make creative exhibits that will keep the attention of visitors of all ages.

Lanna Jackson visited Chicago's Museum of Science and Industry with her family a couple times a year when she was in grade school, and later, in junior high, she often came with friends on the weekends. Eventually Lanna and two of her friends decided to join one of the science clubs. The club met every Saturday and conducted a variety of experiments which were determined by the fifteen club members and supervised by two adults. In June the current club ended festivities with the annual Science Club Network Jamboree, where other clubs from museums, schools, and community groups met for a day of science-related competitions. Lanna says that she made so many new friends and learned so much more that she plans on joining the club next year, too.

INTERNSHIPS

High school is not too soon to start looking for work related to engineering. One of the best ways to locate valuable, hands-on experience is to work as an intern, either after school, on the weekends, or during the summer. To be perfectly honest, there are many more internships available for college students. That shouldn't prevent you from trying to find a local company, museum, or manufacturer who would be willing to exchange experience for a little extra help. Start with the obvious places. Ask your science and math teachers if they know of any companies in the area that offer internships or might be interested in having another pair of eyes, ears, and hands to help with projects. Contact adults you, your parents, or teachers know who are engineers. Where do they work? Could their company or workplace use your help? Be as enterprising as possible. If these leads vanish into thin air, move on to less obvious places. For example, look through the summaries of the chapters included in this book. What is the basic role of each engineer? Focus on the verbs used to describe their work. Transportation engineers, for instance, coordinate many different aspects of traffic. What other companies or workplaces are involved with the coordination of events? How about the mayor's Special Events Office? Or the

Philanthropy Committee for the local college? The scheduling office at an arena or convention center? You get the idea. Brainstorm all of the different possibilities, then ask your parents and friends for their ideas and suggestions. The key to learning always lies in how well you can apply your knowledge. Where you intern and what the company produces are less important than learning how to apply the basic, elemental aspects of engineering in a real-world scenario. In the end, your initiative and creative thinking will go miles toward proving to your next employer what an asset you will be.

EMPLOYMENT

Take your internship experiences to the next level and look for paid jobs that not only use your blossoming engineering experience but also provide you with the opportunity to add to your existing knowledge. Challenge yourself. Start by contacting the same firms you used to search for internships. Now that you have some practical experience under your belt, they may be more inclined to offer you an internship or, better yet, a paying position.

Don't think small, but be realistic. As hard as it is to get internships as a high school student, it's even more difficult to get jobs. If you're finding that employers are saving those choice positions for college students, aim just a little bit lower. Let employers know how interested you are in learning from the ground up and you just might find yourself learning how to operate a tool-and-die machine—and taking home a nice paycheck.

Technician-level jobs, or positions as apprentices to technicians are another good place to look for work. Keep in mind that you will be competing with men and women who have trained in technical schools for some of these jobs, but with solid grades and an obvious desire to learn, your chances are good.

If you still turn up empty-handed, resort to the same tactics that helped you locate more unusual opportunities for internships. Brainstorm the specific tasks which different engineers complete. Mechanical engineers work with moving parts, engines, machine repair. Try calling repair and automotive shops, and the service departments of companies or organizations which use equipment; inquire if they have any openings in maintenance or service/repair. Hospitals, golf courses, apartment buildings, and factories, for instance, all use specific types of machines and devices, and these places all need someone to help repair and maintain the equipment. You might find you have to work for free for a probationary period until you learn the ropes, but in the end, you'll have a paying job that will give you valuable experience.

||| Surf the Web

FIRST

You must use the Internet to do research, to find out, to explore. The Internet is the closest you'll get to what's happening now all around the world. This chapter gets you started with an annotated list of Web sites related to engineering. Try a few. Follow the links. Maybe even venture as far as asking questions in a chat room. The more you read about and interact with engineering personnel, the better prepared you'll be when you're old enough to participate as a professional.

One caveat: you probably already know that URLs change all the time. If a Web address listed below is out of date, try searching on the site's name or other key words. Chances are, if it's still out there, you'll find it. If it's not, maybe you'll find something better!

THE LIST

ASK DR. SCIENCE . . . HE KNOWS MORE THAN YOU DO!
http://www.ducksbreath.com/

"There is a thin line between ignorance and arrogance," writes Dr. Science, "and only I have managed to erase that line." If you're a fan of his daily radio show, you now have a chance to get another dose of this sly, self-spoofing comedian-scientist on-line.

Click on the Question of the Day—and remember not to take it too seriously. One week's worth of questions involved computer vents, cartoon lip reading, women and lingerie, the speed of light in Montana, and programming in the dark. This site is more than just a radio program typed on the screen; it offers honest-to-goodness interactivity. For instance, you can submit your own question, or visit the vault of knowledge to search by keyword for any topic that you think Dr. Science might have covered. (Let your mind run wild.)

If you like this site so much you never want to miss it, you can ask Dr. Science to send you his daily Q&A by email. Real science-joke junkies can ask for a printed catalog of paraphernalia and bulletins.

BOOKER T. WASHINGTON HIGH SCHOOL FOR THE ENGINEERING PROFESSIONS
http://www.hal-pc.org/~fdw/

This Texas high school within a high school is for students who already know they want to specialize in math, science, engineering, and computer science. If you're drawn to the idea of jumping neck-deep into engineering at an early age, you'll want to take a look at this site.

What the High School for Engineering Professions (HSEP) offers is essentially a college-prep program loaded with all-honors courses. Go straight to the section called Frequently Asked Questions, because you've probably got them. The application process is all explained here. If you apply and are among the one hundred new students accepted each year, you'll enroll in rigorous classes in computer science, engineering, engineering graphics, technical writing, and other subjects that your friends back home will be impressed by.

The goal of HSEP's students, of course, is to have great colleges salivating over them when they graduate. In 1996 the seventy-five graduating students received a total of over three million dollars in scholarship offers. HSEP also boasts of close ties with NASA, Rice University, and Exxon, all of whom provide summer employment and internship opportunities.

If you're enticed, there's a place to email for more information. According to the Web site, there are students coming to HSEP from all around the world. But even if you don't live in (or want to move to) Texas, this site is food for thought, as there are numerous other science and engineering high schools you can search for on the Internet.

BRIDGE ENGINEERING
http://www.best.com/~solvers/bridge.html

Are you in awe of those stretches of bridge that seem to be held up by mere wires or planks of wood? You're not alone. In fact, there's a whole engineering community devoted to building and studying bridges. Read here about bridge "events" in places as far away as Edinburgh, Scotland, and Mumbai, India—and as close to home as Columbus, Ohio.

The site was created by SC Solutions, a California engineering firm, in order to provide a comprehensive link to all bridge-related material on the Web. Even if you think a bridge is a bridge is a bridge, this site could make you think again. There are descriptions of prestressed concrete bridges, covered bridges, wood bridges, and so on, all with societies or councils you can link to from here. Many federal and state transportation agencies are also represented by links, as are a number of universities and research institutions. If you're scouting out internship possibilities, you might look at the list of links to U.S. companies that do bridge construction.

The Exhibits section is a wonderful diversion—even for casual bridge enthusiasts. You can link to photos of a popsicle stick bridge, the world's largest bridges, or view a video clip about the Tacoma Narrows Bridge, one of the most documented bridge failures ever.

ENGINEERING: YOUR FUTURE

http://www.asee.org/precollege

Sponsored by the American Society for Engineering Education, this Web site supplies exactly the information you "precollege engineers" are looking for. It's presented in a great question-and-answer format, with questions that don't insult your intelligence. They touch on a range of topics, including the diversity of jobs within engineering, the high school courses you'll need to get accepted into a good engineering institution, and the SAT or ACT scores you should shoot for.

In case you're wondering whether you'll be able to handle those engineering courses in college, you can take a science/math aptitude test on-line at this site. If you make it over that hurdle, there's specific information to help you pick the right engineering school and secrets on getting admitted. Then link right to the Web pages of hundreds of colleges and universities in the United States that offer strong engineering programs.

Of course, you'll need to pay for college, and this site has considerately thought of that too. There's a dollar-wise section with links to the Web sites of Department of Education and various federal loan, grant, and work-study programs.

EXPLORING YOUR FUTURE IN MATH AND SCIENCE

http://www.cs.wisc.edu/~karavan/afl/home.html

This site is the culmination of a project conducted for a women's studies course at the University of Wisconsin-Madison. It tackles a subject you've probably heard about in the media—the fact that while girls enjoy math and science classes in early grades, they tend to shy away from the subjects when they reach adolescence.

This site explores ways that girls can revive their interest in science and math, particularly as a means of exploring broader career options. A page titled "How much money is in it for me?" shows the salary for a junior engineer starting at $34,000, with chief engineers earning $80,000. Another page discusses the value of high school science clubs and regional competitions and encourages girls to participate—or better yet, start their own science group. Yet another page offers girls a lesson in critical thinking. It suggests that students should analyze the story problems used in their math books for what kind of activities women and girls, as opposed to men and boys, are engaged in.

This site—while intriguing—doesn't provide a conclusive or comprehensive discussion of the subject. However, if it whets your appetite, you'll find a slim list of links to other Internet resources here, such as the Society of Women Engineers, the Association for Women in Science, and the Archives of Women in Science and Engineering.

HIGH SCHOOL CHEMISTRY 250+ LINKS
http://www.home.ptd.net/~swenger/

No matter what engineering specialty you plan on choosing as a major, you're going to have to make it through high school chemistry first . . . and get the kind of grades that make you stand out.

The Webmaster of this site is a high school chemistry teacher with a heart of gold and thirty-five years of teaching under his belt. He decided to find and bring together resources appropriate for high school chemistry students and to organize them in a useful fashion. That's no small task.

What you'll find are links to over 250 sites, organized by topics such as nuclear chemistry, organic chemistry, chemistry competitions, periodic tables, famous scientists, etc. If you don't find what you're looking for, you can even email the author with suggestions, recommendations, or questions.

HUMAN-POWERED HYDROFOIL
http://lancet.mit.edu/decavitator/

It's a bird, it's a plane . . . it's a Decavitator. A what? If you're worried that college is going to be all work and no play, check out this site for a blend of the two. The students at the Massachusetts Institute of Technology (MIT) are known for their engineering prowess, and the Decavitator, a human-powered hydrofoil, is yet another demonstration of it. This student-developed boat was funded by MIT's SeaGrant program, and in 1993 was even awarded the DuPont prize for the fastest human-powered watercraft. Each year, students at MIT work to improve its performance.

Visit this site to read a description of hydrofoil basics that even a literature major could understand, and then watch videos of the Decavitator in action. If you're inspired to build one of your very own, you'll find a three-dimensional drawing and specifications, as well as a page that describes methods and materials used to manufacture the boat. And, should you *really* want to talk hydrofoil, you can even email some of the members of the team who built it.

IMAGINE

http://jhunix.hcf.jhu.edu/%7Esetmentr/Imagine.html

Imagine is a bimonthly journal for the go-getter high school student with his or her eye on the future. Its tag line, "Opportunities and resources for academically talented youth," says it all.

If you're always searching for good academic programs, competitions, and internships, this publication can keep you well informed on what's available and when you need to apply. There's an entertaining College Review series in which student contributors evaluate individual colleges and universities and also a Career Options series featuring interviews with professionals.

Along with the current issue, selected portions of back issues can be read on-line. Previous issues have included articles about the USA Computing Olympiad and the ThinkQuest Competition, as well as general tips on entering academic competitions and choosing summer academic programs. For $30 you can subscribe and get the printed journal delivered to your home—or for free, you can just read back issues on-line.

JUNIOR ENGINEERING TECHNICAL SOCIETY (JETS)

http://www.asee.org/jets/

Here's another spin-off site of the American Society for Engineering Education, and again, there's an unbelievable amount of good stuff. The mission of JETS is to provide opportunities for students to "try out" careers while they're in high school so that they can make informed academic and career choices.

At the JETS home page, you'll see icons identifying three of its major programs. Click on TEAMS (Tests of Engineering Aptitude, Mathematics, and Science) to learn about this program in which teams of students work with an engineering mentor, then participate in an open-book, open-discussion engineering problem competition. The National Engineering Design Challenge is another program that challenges team of students working with an advisor to design, fabricate, and demonstrate a working solution to a social need. Finally, take the self-administered National Engineering Aptitude Search to determine if your education is on the right track.

For each of these programs, you can view sample problems and their solutions. In addition, many brochures are posted on-line, including the excellent series "Engineering and You," which provides descriptions of over twenty disciplines of engineering (plastics, electrical, nuclear, optical, etc.). If you'd like to have something to hold in your very own hands, JETS will also send you (that is, through the mail) a broad range of brochures, books, and videos.

OMNISCIENCE FUTURENEERING

http://www.webcom.com/sknkwrks/

Omniscience Futureneering is a high school club in Florida founded on the premise that the best way to learn about something is to "do it, make it, or use it." Each year this quirky group devotes its resources to a major science project that intrigues them—the only criteria being that the project somehow involve the future.

There aren't any deep-pocketed corporate sponsors footing the bill for Omniscience, and as a result, you could actually do many of their projects yourself. For instance, rather than buy a $40 kit to make a test-tube Stirling engine, they created one from a Pyrex test tube, cloth tape, paper clips, marbles, the end of a balloon, steel wool, and other supplies.

The voice and pillar behind Omniscience Futureneering is a high school teacher who writes that his teaching goal is to "pass on the basics of physics so that my students are science-literate and able in turn to pass on that knowledge." He reveals that his greatest sorrow is "that popular mechanical and science magazines no longer have the make it, do it, try it, cut yourself, electrocute yourself, burn yourself stories."

Okay, so there's an aura of mad scientist about this site. Nonetheless, you'll enjoy reading the lengthy but fascinating descriptions of the projects—like the radio-controlled, TV-eye car—and learning from their flops and successes. For the cost of postage, they'll even draw up sketches or print photos of a project's components and send them to you.

PETERSON'S GUIDE TO SUMMER PROGRAMS FOR TEENAGERS

http://www.petersons.com/summerop/

Your commitment to a brilliant academic future might waver when you visit this site. Along with some great information about academic and career-focused summer programs, you'll be tantalized by summer camps that revolve around activities that are less mentally rigorous—like white water rafting or touring Switzerland on a bicycle. Shake it off. You're here to further your education, and this site offers good tips on assessing any summer program or camp you're considering.

Finding a camp that suits your interests is easy enough at this site—just search Peterson's database of academic, travel, and camping programs. Type in the keyword "science," for instance, and you'll bring up a list of links to over two hundred summer programs, including ScienceQuest, Cascade Science School, Yale Summer Programs, and Smith College Summer Science Program. Then click on a specific program or camp for a quick overview description. In some instances you'll get a more in-depth description, along with photographs, applications, and on-line brochures. If you need to limit your search to your home state, that's easy enough, too. You can sift through Peterson's database by geographic region or alphabetically.

TRACNET

http://www.trac.net/

Future civil engineers will definitely find brain candy at this site. TRAC's mission is to introduce high school students to the field of transportation through a unique math and science curriculum. The real meat of this site, though, can be found in a section called Student Resources, which was created just for students considering a college degree in civil/transportation engineering.

Within this colorful, well-designed section, TRAC has compiled a list of universities and colleges with strong degree programs in civil engineering and sorted them by region and state. This could save you a ton of time doing your own research. To help you find gainful summer employment, they've also built links to several summer research programs and job centers, as well as to every state's Department of Transportation (many of which offer internships).

While you're here you'll also want to explore areas within Student Resources that are dedicated to professional organizations and high school education requirements. Other promising finds include Career Link and Mentoring Connection.

U.S. News & World Report Career Guide: Engineering

http://www.usnews.com/usnews/edu/BEYOND/BCENG.HTM

Want pragmatic tidbits on the specialty of engineering that looks most promising? This *U.S. News & World Report* Career Guide site will help you. The trick, as the editorial here points out, is in picking a specialty that's "hot" at the time you graduate and are ready to start working.

The site describes the current market for engineers, which is booming. Electrical engineers account for one-quarter of the total engineers currently working now, and that demand is expected to stay high. With computers permeating every aspect of life, computer engineering (a relatively new specialty) is expected to grow. Biomedical engineering is another specialty that appears to be on the rise for the future—or maybe not. You can read all about the pros and cons of this specialty.

For nuts and bolts info for future engineers, this site hosts several great search engines. You can search the Occupational Outlook Handbook by keyword or search for information about an engineering school. It's also possible that you're still dabbling with the idea of careers other than engineering—like journalism, forestry, political science, musicology (after all, the world is your oyster). If that's the case, hunker down in front of your computer screen and enjoy all that this expansive career site has to offer.

Yahoo: Engineering

http://www.yahoo.com/Science/Engineering/

It might seem odd to include the popular search engine Yahoo among a list of engineering Web sites, but it won't seem so after you've visited it. If you're hungry for more after visiting the sites listed in this appendix, pull up a chair at Yahoo's feast.

Yahoo has done a tremendous amount of legwork for you. For example, is it electrical engineering you're interested in? Then, scan through the 1,051 sites currently included here. Industrial engineering posts an impressive 124 sites. Even welding engineering offers 14 places you probably wouldn't have known to look for otherwise. There are oodles of other engineering professions listed here, as well as engineering education sites, events, journals, and magazines.

Read a Book

FIRST

When it comes to finding out about engineering, don't overlook a book. (You're reading one now, after all.) What follows is a short, annotated list of books and periodicals related to engineering. The books range from fiction to personal accounts of what it's like to be an engineer, to professional volumes on specific topics. Don't be afraid to check out the professional journals, either. The technical stuff may be way above your head right now, but if you take the time to become familiar with one or two, you're bound to pick up some of what is important to engineers, not to mention begin to feel like a part of their world, which is what you're interested in, right?

We've tried to include recent materials as well as old favorites. Always check for the latest editions, and, if you find an author you like, ask your librarian to help you find more. Keep reading good books!

BOOKS

Baldwin, Neil. Edison: *Inventing the Century*. New York: Hyperion, 1996. A revealing portrait of one of America's most significant inventors, capturing his imagination, dynamism, and entrepreneurial brilliance.

Bilstein, Roger E. *The American Aerospace Industry: From Workshop to Global Enterprise. Twayne's Evolution of Modern Business*. New York: Twayne Publishers, 1996. Detailed coverage of various aerospace industries, the history of aerospace in the United States, the aircraft industry, and much more.

Brandt, Daniel A. *Metallurgy Fundamentals*. South Holland: Goodheart-Willcox Co., 1992. Engaging overview of metallurgy theory and practice.

Bucciarelli, Louis L. *Designing Engineers: Inside Technology.* Cambridge, MA: MIT Press, 1996. An engineer and educator answers the questions of how the products we use every day—products of engineering design—came to be the way they are, from conceptualization to production.

Bulger, Ruth Ellen, Elizabeth Meyer Bobby, and Harvey V. Fineberg. *Society's Choices: Social and Ethical Decision Making in Biomedicine.* Washington, DC: National Academy Press, 1995. Presents a fascinating set of perspectives on ethical issues in biomedicine and biomedical engineering, incorporating the recommendations of various interest groups.

Chaikin, Andrew. *Air and Space: The National Air and Space Museum Story and Flight.* Boston: Bullfinch Press, 1997. Drawing on the peerless collection of the National Air and Space Museum, this magnificent book captures the daring innovations and adventures that became a part of humanity's conquest of the skies.

Doherty, Paul. *Cyberplaces: The Internet Guide for Architects.* Duxbury: Robert Means Co., 1997. From the only well-known authority on the subject, the only available guide to the application of the Internet and related technology to architecture, contracting, and facility management.

Dubbel, Heinrich, W. Beitz, and K. H. Kuttner. *Handbook of Mechanical Engineering.* New York: Springer Verlag, 1994. An excellent manual to mechanical engineering, technology, and the industrial arts.

Fant, Kenne. *Alfred Nobel: A Biography.* New York: Arcade Publishers, 1993. A biography of the inventor of dynamite and founder of the Nobel Prize. Compelling account of a complicated and troubled man.

Fenichell, Stephen. *Plastic: The Making of a Synthetic Century.* New York: HarperBusiness, 1997. An expert and entertaining look at plastic in the twentieth century—its social history and cultural legacy.

Gilbreth, Lillian Moller. *The Quest of the One Best Way: A Sketch of the Life of Frank Bunker Gilbreth.* New York: Society of Women Engineers, 1990. An in-depth and fascinating look at a great industrial and transportation engineer.

Goetsch, David L. *Occupational Safety and Health in the Age of High Technology: For Technologists, Engineers, and Managers.* 3rd ed. Paramus: Prentice Hall, 1996. Excellent book that addresses the need for a practical and informed teaching resource focusing on the needs of modern health and safety professionals in today's workplace.

Goldberg, David E. *Life Skills and Leadership for Engineers*. New York: McGraw Hill Text, 1995. Discusses the major skills and qualifications needed to thrive in the engineering world, focusing on the importance of leadership and responsibility.

Hambley, Allan R. *Electrical Engineering: Principles and Applications*. Paramus: Prentice Hall, 1997. User-friendly text providing a solid foundation in the basics of circuits, electronics (radio and digital), and electro-mechanics.

Hawkins, Lori, and Betsy Dowling. *100 Jobs in Technology*. New York: MacMillan General Reference, 1997. Among the many careers and jobs listed in this comprehensive book are: CD-ROM producer, environmental engineer, physicist, internet access provider, and biotechnology researcher.

Heaton, Alan, ed. *The Chemical Industry*. 2nd ed. New York: Van Nostrand Reinhold, 1994. An excellent overview of the field, designed for undergraduate students of chemistry and chemical engineering.

Hosler, Dorothy. *The Sounds and Colors of Power: The Sacred Metallurgical Technology of Ancient West Mexico*. Cambridge: MIT Press, 1994. An interesting scholarly account of early methods of materials science, tracing the roots of metallurgy back to a West Mexican technological civilization.

Jarvis, Adrian. *The Liverpool Dock Engineers*. Littlehampton: Sutton, 1997. First book to be written specifically on dock engineers and their work; discusses in great detail the development of the field and its working methods.

Kletz, Trevor. *What Went Wrong?: Case Histories of Process Plant Disasters*. 3rd ed. Houston: Gulf Publishing Co., 1994. An insightful examination of various case histories that reveals the causes and aftermaths of various plant disasters, providing insight into how such catastrophes can be avoided.

Lampl, Richard. *The Aviation & Aerospace Almanac 1998*. New York: McGraw Hill, 1997. Filled to the brim with interesting information and facts about space engineering, science, mathematics, transportation, and general aviation.

Lewis, Grace Ross. *1,001 Chemicals in Everyday Products*. New York: Van Nostrand Reinhold, 1994. Informative and user-friendly guidebook covering a wide range of chemicals in common household products, from food, cosmetics, and cleaning products to gardening and care products.

Lewis, Tom. *Divided Highways: Building the Interstate Highways, Transforming American Life*. New York: Viking Press, 1997. An informative and lively account of our greatest public works project—a rich brew of energy, greed, and dreams.

Maples, Wallace. *Opportunities in Aerospace Careers*. VGM Opportunities. Lincolnwood: NTC Publishing Group, 1995. Offers excellent vocational advice to the young reader about various fields within aerospace, especially aeronautical engineering.

Meikle, Jeffrey L. *American Plastic: A Cultural History*. New Brunswick: Rutgers University Press, 1995. An illustrated and detailed exploration of America's love-hate relationship with plastic, from Bakelite radios and nylon stockings to Tupperware and Disneyworld.

Miller, Rex, and Mark R. Miller. *Electronics the Easy Way*. 3rd ed. Hauppauge: Barron's Educational Series, 1995. A useful and easy-to-read guide for anyone interested in electronics.

Nalder, Eric. *Tankers Full of Trouble: The Perilous Journey of Alaskan Crude*. New York: Grove Press, 1994. An investigative journalist offers a thrilling, life-or-death account of an oil tanker—the dangers of ship and sea, the everyday distractions, and the extremes of weather and fatigue.

Peterson's Guides. *Peterson's Graduate Programs in Engineering & Applied Sciences 1997*. 31st ed. Princeton: Peterson's, 1997. Lists more than 3,400 programs, from bioengineering and computer science to mechanical and electrical engineering. Absolutely indispensable.

Petroski, Henry. *Engineers of Dreams: Great Bridge Builders and the Spanning of America*. New York: Vintage Books, 1996. A great writer and historian reveals the science, the politics, the egotism, and the magic of America's great bridges, recounting the fascinating stories of the men and women who built them.

Pollard, Michael. *The Lightbulb and How It Changed the World. History and Invention*. New York: Facts on File, 1995. Takes us on a sweeping tour of the light bulb's history, discussing along the way its social impact on education, leisure, and work.

Richardson, Terry L., and Erik Lokensgard. *Industrial Plastics: Theory and Application*. 3rd ed. Albany: Delmar Publications, 1997. An informative and readable technical account of plastics and plastics engineering.

Robertson, H. Douglas, and Joseph E. Hummer. *Manual of Transportation Engineering Studies*. Paramus: Prentice Hall, 1994. Shows in detail how to conduct transportation engineering studies in the field—beginning with basic studies and moving to complex or specialized studies.

Rooney, M. L. *Active Food Packing*. New York: Chapman & Hall, 1995. Provides an overview of packaging active foods, discussing produce, packaging agents, bottling, safety considerations, and more.

Rubin, Irvin I. *Handbook of Plastic Materials and Technology*. New York: Wiley-Interscience, 1990. Comprises 119 chapters on plastic materials, properties, processes, and industry practices.

Seifer, Marc J. *Wizard: The Life and Times of Nikola Tesla: Biography of a Genius*. Secaucus: Birch Lane Press, 1996. An excellent and comprehensive biography of one of the most influential and enigmatic electrical engineers of our century.

Selke, Susan E. *Packaging and the Environment: Alternatives, Trends and Solutions*. Lancaster: Technomic Publishing Co., 1994. Examines resource use, pollution, solid waste, landfills, reduction and reuse, recovery and incineration, composting, recycling, degradable packaging, and more.

Shames, Irving H. *Engineering Mechanics: Dynamics*. 4th ed. Paramus: Prentice Hall, 1997. The latest edition of a highly respected and well-known text for courses in engineering mechanics. Accessibly focused on fundamental principles instead of "cookbook" problem solving.

Sparke, Penny. *The Plastics Age: From Bakelite to Beanbags and Beyond*. New York: Overlook Press, 1994. A history of plastics products from the 19th century to the present, in 15 essays and 150 color and black-and-white photographs. Explores such topics as industrial design and commercial art, perceptions of plastic, and pop culture.

Strong, A. Brent. *Plastics: Materials and Processing*. New York: Prentice Hall, 1996. Introduces plastic to a wide range of readers who either need to gain, refresh, or improve their knowledge of plastic materials and processing.

White, John H. *The American Railroad Freight Car: From the Wood-Car Era to the Coming of Steel*. Baltimore: Johns Hopkins University Press, 1995. A history of the American freight car, discussing the various types of specialized cars used to handle America's growing freight traffic, and examining the technological developments that influenced freight-car design.

White, Richard M., and Roger Doering. *Electrical Engineering Uncovered.* Paramus: Prentice Hall, 1996. Perfect undergraduate introduction to electrical engineering that also gives a good sense of what professional engineers do.

Wood, Robert B., and Kenneth R. Edwards. *Opportunities in Electrical Trades.* VGM Opportunities Series. Lincolnwood: NTC Publishing Group, 1997. Offers sound vocational guidance for aspiring electrical engineers, with a focus on emerging careers.

PERIODICALS

Aerospace Engineering Magazine. Published monthly by the Society of Automotive Engineers, Inc., 400 Commonwealth Drive, Warrendale, PA 15096-0001. A detailed and nicely produced survey of breakthroughs in the aerospace trade, published by a premier organization in automotive and aerospace engineering.

American Society of Highway Engineers Scanner Newsletter. Updated daily by the American Society of Highway Engineers at http://www.highwayengineers.com. Extensive coverage of developments—state, federal, or otherwise—in highway transportation. Offers an excellent sense of the complexity of public projects.

AMM On-line. Updated twice daily by the American Metal Market at http://www.amm.com. A colorful and thorough news network for metallurgists, offering top stories about the metals trade all over the world.

Annals of Biomedical Engineering. Published bimonthly by the Department of Biomedical Engineering, Harris Hydraulics Lab, Room 310, University of Washington, Box 357962, Seattle, WA 98195-7962. Interdisciplinary, international journal collecting original articles in major fields of bioengineering and biomedical engineering.

ASME International. Published monthly by the American Society of Mechanical Engineers, 345 East 47th Street, New York, NY 10017. Covers aerospace, computers in engineering, design engineering, electronic packaging, environmental control, and other issues that pertain to mechanical engineering.

Chemical Engineering. Published monthly by The McGraw-Hill Companies, 1221 Avenue of the Americas, New York, NY 10020. Offers indispensable and entertaining coverage of the industry: up-to-date articles on new products and the companies that make them, the job field, safety measures, and technological innovations.

Chemical Week. Published weekly by Chemical Week Associates, 888 Seventh Avenue, 26th Floor, New York, NY 10106. A fun and lively magazine about the diverse chemical industry, offering news and in-depth coverage of the theory, practice, and business of chemicals.

Converting Magazine. Published monthly by Reed Elsevier Inc., 275 Washington Street, Newton, MA 02518-1630. Official publication for manufacturers involved in converting paper, paperboard, film, and foil into packaging and other products.

Electronic Engineering Times On-line. Updated daily at http://techweb.cmp.com. An excellent guide to the industry that offers day-to-day news, career advice, product analysis, features, and columns.

Electronic Products: The Engineer's Magazine of Product Technology. Published monthly by Hearst Business Communications Inc., Electronic Products, 645 Stewart Avenue, Garden City, NY 11530. Details both the latest developments in electronics and unusual or new applications of products or technologies.

International Journal of Industrial Engineering. Published quarterly by the School of Industrial Engineering, University of Cincinnati, Cincinnati, OH 45221-0116. With an emphasis on engineering design, covers consumer product design, engineering economy and cost estimation, information systems, materials handling, and much more.

ITE Journal. Published quarterly by the Institute of Transportation Engineers, 525 School Street, SW, #410, Washington, DC 20024-2797. Presents research in transportation planning, geometric design, traffic operations, goods movement, signs and markings, parking, safety, ride sharing, and mass transit.

Journal of Clinical Engineering. Published quarterly by The Journal of Clinical Engineering, 1185 Avenue of the Americas, New York, NY 10036. Focuses on the needs of hospital-based clinical engineers, educators, researchers, and other professionals engaged with the development and applications of medical technology.

Journal of Manufacturing Science and Engineering. Published monthly by the American Society of Mechanical Engineers, 345 East 47th Street, New York, NY 10017. Articles on computer-integrated manufacturing, design for manufacturing, girding and abrasive machining, inspection and quality control, robotics, and a lot more.

Journal of Technology Education. Published quarterly by the Journal of Technology Education, Virginia Tech University, 144 Smyth Hall, Blacksburg, VA 24601-0432. Provides a forum for scholarly discussion of technology education research, philosophy, theory, and practice. Book reviews, articles, and interviews.

Packaging Digest. Published monthly by Reed Elsevier Inc., 275 Washington Street, Newton, MA 02518-1630. Thorough review of the packaging industry, specializing in food and electronics packaging.

Plastics Engineering. Published monthly by the Society of Plastics Engineers, Inc., PO Box 403, Brookfield, CT 06804-0403. Explores, through technical articles and industry news, the science and technology of plastics engineering within the context of the plastics business.

Ask for Money

By the time most students get around to thinking about applying for scholarships, they have already extolled their personal and academic virtues to such lengths in essays and interviews for college applications that even their own grandmothers wouldn't recognize them. The thought of filling out yet another application form fills students with dread. And why bother? Won't the same five or six kids who have been fighting over grade point averages since the fifth grade walk away with all the really *good* scholarships?

The truth is, most of the scholarships available to high school and college students are being offered because an organization wants to promote interest in a particular field, encourage more students to become qualified to enter it, and finally, to help those students afford an education. Certainly, having a good grade point average is a valuable asset, and many organizations that grant scholarships request that only applicants with a minimum grade point average apply. More often than not, however, grade point averages aren't even mentioned; the focus is on the area of interest and what a student has done to distinguish himself or herself in that area. In fact, sometimes the *only* requirement is that the scholarship applicant must be studying in a particular area.

GUIDELINES

When applying for scholarships there are a few simple guidelines that can help ease the process considerably.

Plan Ahead

The absolute worst thing you can do is wait until the last minute. For one thing, obtaining recommendations or other supporting data in time to meet an application deadline is incredibly difficult. For another, no one does their best thinking or writing under the gun. So get off to a good start by reviewing scholarship applications as early as possible—months, even a year, in advance. If the current scholarship information isn't available, ask for a copy of last year's version. Once you have the scholarship information or application in hand, give it a thorough read. Try to determine how your experience or situation best fits into the scholarship, or if it even fits at all. Don't waste your time applying for a scholarship in literature if you couldn't finish *Great Expectations.*

If possible, research the award or scholarship, including past recipients and, where applicable, the person in whose name the scholarship is offered. Often, scholarships are established to memorialize an individual who majored in religious studies or who loved history, for example, but in other cases, the scholarship is to memorialize the *work* of an individual. In those cases, try to get a feel for the spirit of the person's work. If you have any similar interests or experiences, don't hesitate to mention these.

Talk to others who received the scholarship, or to students currently studying in the same area or field of interest in which the scholarship is offered, and try to gain insight into possible applications or work related to that field. When you're working on the essay asking why you want this scholarship, you'll have real answers—"I would benefit from receiving this scholarship because studying engineering will help me to design inexpensive but attractive and structurally sound urban housing."

Take your time writing the essays. Make sure you are answering the question or questions on the application and not merely restating facts about yourself. Don't be afraid to get creative; try to imagine what you would think of if you had to sift through hundreds of applications: What would you want to know about the candidate? What would convince you that someone was deserving of the scholarship? Work through several drafts and have someone whose advice you respect—a parent, teacher, or guidance counselor—review the essay for grammar and content.

Finally, if you know in advance which scholarships you want to apply for, there might still be time to stack the deck in your favor by getting an internship, volunteering, or working part-time. Bottom line: the more you know about a scholarship and the sooner you learn it, the better.

Follow Directions

Think of it this way: many of the organizations that offer scholarships devote 99.9 percent of their time to something other than the scholarship for which you are applying. Don't make a nuisance of yourself by pestering them for information. Simply follow the directions as they are presented to you. If the scholarship application specifies that you write for further information, then write for it—don't call.

Pay close attention to whether you're applying for an award, a scholarship, a prize, or financial aid. Often these words are used interchangeably, but just as often they have different meanings. An award is usually given for something you have done: built a park or helped distribute meals to the elderly; or something you have created: a design, an essay, a short film, a screenplay, or an invention. On the other hand, a scholarship is frequently a renewable sum of money that is given to a person to help defray the costs of college. Scholarships are given to candidates who meet the necessary criteria based on essays, eligibility, grades, or sometimes all three.

Supply all the necessary documents, information, and fees, and make the deadlines. You won't win any scholarships by forgetting to include a recommendation from a teacher or failing to postmark the application by the deadline. Bottom line: get it right the first time, on time.

Apply Early

Once you have the application in hand, don't dawdle. If you've requested it far enough in advance, there shouldn't be any reason for you not to turn it well in advance of the deadline. You never know, if it comes down to two candidates, your timeliness just might be the deciding factor. Bottom line: don't wait, don't hesitate.

Be Yourself

Don't make promises you can't keep. There are plenty of hefty scholarships available, but if they all require you to study something that you don't enjoy, you'll be miserable in college. And the side effects from switching majors after you've accepted a scholarship could be even worse. Bottom line: be yourself.

Don't Limit Yourself

There are many sources for scholarships, beginning with your guidance counselor and ending with the Internet. All of the search engines have education categories. Start there and search by keywords, such as "financial aid," "scholarship," and "award." But don't be limited to the scholarships listed in these pages.

If you know of an organization related to or involved with the field of your choice, write a letter asking if they offer scholarships. If they don't offer scholarships, don't stop there. Write them another letter, or better yet, schedule a meeting with the president or someone in the public relations office and ask them if they would be willing to sponsor a scholarship for you. Of course, you'll need to prepare yourself well for such a meeting because you're selling a price-less commodity—yourself. Don't be shy, be confident. Tell them all about your-self, what you want to study and why, and let them know what you would be willing to do in exchange—volunteer at their favorite charity, write up reports on your progress in school, or work part-time on school breaks, full-time dur-ing the summer. Explain why you're a wise investment. Bottom line: the sky's the limit.

THE LIST

AHS Vertical Flight Foundation Scholarships
American Helicopter Society
217 North Washington Street
Alexandria, VA 22314
Tel: 703-684-6777

Thirteen applicants will receive $2,000 awards. Included in the application process: demonstration of character, recommendations, personal essay, and transcript.

Air Force ROTC Scholarships
Scholarship Actions Branch
551 East Maxwell Boulevard
Maxwell AFB, AL 36112
Tel: 800-522-0033, ext. 2825

Four-year scholarships are available to high school students planning to study engineering, architecture, computer science, or physics in college.

American Council for Construction Education Scholarships
901 Hudson Lane
Monroe, LA 71201
Tel: 318-323-2413

Undergraduate scholarships of up to $1,500 per year and $7,500 graduate fel-lowships are offered for study in civil engineering or construction.

American Institute of Aeronautics and Astronautics Scholarships
370 L'Enfant Plaza
Washington, DC 20024
Tel: 202-646-7400

Students who have completed at least one term in college and plan to work as an engineer or scientist in the field may apply for $1,000 scholarships.

American Radio Relay League Foundation Scholarships
225 Main Street
Newington, CT 06111
Tel: 860-666-1541

Members of the league who are majoring in electrical engineering are eligible for scholarship assistance. Awards range from $500 to $1,000.

American Society of Civil Engineers
345 East 47th Street
New York, NY 10017
Tel: 212-705-7514

Members of a campus chapter of ASCE may apply for $1,000 to $2,000 scholarships to help finance study in civil engineering; ASCE also offers awards to support student research. Apply by February 1.

Archie Memorial Scholarship
Society of Petroleum Engineers International
Professional Development Department
PO Box 833836
Richardson, TX 75083
Tel: 214-952-9393

This $3,000 scholarship is available to first-year college students majoring in petroleum engineering. You may also write for a list of local chapters that may award aid.

ASM Foundation Undergraduate Scholarships
ASM International
Metals Park, OH 44073
Tel: 216-338-5151

Scholarships are intended for high school or college students who are studying or planning to study metallurgy or materials science.

Associated General Contractors Education and Research Foundation Scholarships
1957 E Street, NW
Washington, DC 20006
Tel: 202-393-2040

College sophomores, juniors, and seniors who are majoring in construction or civil engineering are eligible for scholarships of $1,500 per year.

Association for Manufacturing Technology Scholarships
7901 Westpark Drive
McLean, VA 22102
Tel: 703-893-2900

These scholarships of up to $2,000 per year are available to college freshmen pursuing careers in the machine tool industry. Application forms are available at participating colleges and universities.

Association of Engineering Geologists Scholarships
323 Boston Post Road, Suite 2D
Sudbury, MA 01776
Tel: 508-443-4639

Scholarships of up to $1,000 are offered to members for study in engineering geology.

Bell and Howell Education Group Scholarships
2201 West Howard Street
Evanston, IL 60202
Tel: 800-323-4256

Undergraduates studying computer science or engineering are eligible for scholarship assistance. Write for details. Applications due March 15.

Chemists Club of New York Scholarships
Scholarship Committee
52 East 41st Street
New York, NY 10017
Tel: 212-532-7649

Scholarships are available to student chemists and chemical engineers with strong academic records; write for details.

Cincinnati Milacron Foundation Scholarship Program
4701 Marburg Avenue
Cincinnati, OH 45209

Engineering students may apply for scholarship assistance; applications are due February 1.

▓Civil Air Patrol Scholarships
National Headquarters-TT
Maxwell AFB, AL 36112
Tel: 334-293-5332

Scholarships of $750 are available to high school seniors who are CAP members for undergraduate study in engineering. Write for complete details.

▓Computer Associates Scholarships
One Computer Associates Plaza
Islandia, NY 11788
Tel: 516-342-5224

Undergraduate students in computer science or engineering are eligible for $5,000 scholarships; apply through one of the fifty participating colleges or universities.

▓Construction Education Foundation Scholarships
1300 North 17th Street West
Rosslyn, VA 22209

Students enrolled in a construction curriculum may compete for scholarships ranging from $500 to $2,000. Awards are based on a review of academic record, financial need, extracurricular activities, past employment, and recommendations. Applications are due December 15.

▓Cooper Union for the Advancement of Science and Art Scholarships
41 Cooper Square
New York, NY 10003
Tel: 212-254-6300

High school graduates pursuing study in engineering or architecture are eligible for two hundred renewable, full-tuition scholarships. The deadline to apply is February 1.

▓David Laine Scholarships
North American Die Casting Association
9701 West Higgins Road, Suite 880
Rosemont, IL 60018-4921
Tel: 708-292-3600

Undergraduate or graduate engineering students may apply for these $2,000 scholarships; applicants must intend to pursue a career in the die casting industry. Applications are due August 15.

Edison-McGraw Scholarship Program
Edison Foundation
3000 Book Building
Detroit, MI 48226
Tel: 313-965-1149

This program is open to high school students planning to study science, engineering, or technology. Applicants must submit a proposal on an experiment or project idea; applications are due December 1.

Edward D. Hendrickson/SAE Engineering Scholarship
Society of Automotive Engineers
400 Commonwealth Drive
Warrendale, PA 15096-0001
Tel: 412-772-8534

Applicants must be U.S. citizens who intend to earn a degree in engineering; they must be high school seniors with a GPA of at least 3.75 and rank in the 90th percentile in both mathematics and verbal on the ACT or SAT. This scholarship is $4,000, paid at the rate of $1,000 per year. The scholarship lasts 4 years, provided the recipient maintains a GPA of at least 3.0 in college.

Electronic Industries Foundation Scholarships
919 18th Street, NW
Washington, DC 20006
Tel: 202-955-5810

The foundation awards $2,000 renewable scholarships to undergraduate and graduate students with disabilities. They must study aeronautics, computer science, electrical engineering, engineering technology, applied mathematics, or microbiology. The application deadline falls in February.

Erhardt C. Koerper, P.E. Memorial Scholarship
National Society of Professional Engineers
1420 King Street
Alexandria, VA 22314-2794
Tel: 703-684-2800

To be eligible for this award, applicants must be high school seniors planning to enroll and study engineering at an EAC-ABET accredited school. Applicants must be American citizens. They must have earned at least a 3.0 GPA, rank in the top twenty-five percent of their class, and have achieved certain scores on the ACT or SAT. Applicants should intend to enter the practice of engineering after graduation.

Flour Daniel Engineering Scholarship

Citizens Scholarship Foundation
PO Box 297
St. Peter, MN 56082
Tel: 507-931-1682

This scholarship of $2,000 is awarded annually. Applications are due March 15; write for details.

Foundation Educational Grants

Society of Hispanic Professional Engineers
5400 East Olympic Boulevard, Suite 306
Los Angeles, CA 90022
Tel: 213-888-2080

This scholarship is for undergraduate and graduate students currently studying or planning to study engineering or science. Students must have completed classwork in algebra, trigonometry, geometry, physics and chemistry.

Foundation for Science and the Handicapped Scholarships

1140 Iroquois, Suite 114
Naperville, IL 60563
Tel: 210-333-4600

Students with physical disabilities who are pursuing degrees in science, mathematics, engineering, or medicine are eligible for scholarships.

Fred M. Young, Sr./SAE Engineering Scholarship

Society of Automotive Engineers
400 Commonwealth Drive
Warrendale, PA 15096-0001
Tel: 412-772-8534

Applicants must be U.S. citizens who intend to earn a degree in engineering. They must be a high school senior with a 3.75 GPA or higher and rank in the 90th percentile in both mathematics and verbal on the ACT or SAT. This scholarship is for $4,000, paid at the rate of $1,000 per year for four years provided the recipient maintains a GPA of at least 3.0. The Young Radiator Company established this scholarship in memory of the company's founder.

General Electric Foundation Scholarships

The College Board
45 Columbus Avenue
New York, NY 10023
Tel: 212-713-8000

Community college students majoring in engineering who have a 3.0 GPA are eligible for these scholarships of up to $4,000 to help them complete their degrees.

General Motors Corporation Scholarships
College Relations Program
3044 West Grand Boulevard
Detroit, MI 48202
Tel: 313-556-3565

Scholarships providing full tuition and other benefits are open to engineering majors with a 3.2 GPA or higher. Applications are available at participating institutions.

Helene Overly Scholarships
Women's Transportation Seminar
808 17th Street, NW, Suite 200
Washington, DC 20006
Tel: 202-223-9669

These scholarships of $3,000 help undergraduate and graduate women finance studies in transportation. Apply through your local chapter by February 1.

Institute of Industrial Engineers Scholarships
PO Box 6150
Norcross, GA 30091
Tel: 770-449-0460

Student members majoring in industrial engineering may apply for scholarships; the deadline is November 15.

International Road Federation Fellowships
525 School Street, SW
Washington, DC 20024
Tel: 202-554-2106

Students pursuing a degree program related to highways and highway transportation may apply for these $4,000 fellowships.

International Society for Optical Engineering Scholarships
PO Box 10
Bellingham, WA 98227
Tel: 360-676-3290

Optical engineering majors are eligible for these $500 to $5,000 scholarships; apply by May 1.

John Eager Scholarships
Association for Information and Image Management
1100 Wayne Avenue, Suite 1100
Silver Spring, MD 20910
Tel: 301-587-8202

High school seniors planning to study engineering, physical science, or mathematics in college are eligible for these $5,000 scholarships. The application process requires a 2,500-word essay.

Kodak Scholars Program
Eastman Kodak Company
343 State Street
Rochester, NY 14650
Tel: 716-724-4994

Undergraduate students in science and engineering are eligible for this program, which covers most of their tuition fees. These monetary awards are offered through around twenty-five cooperating colleges and universities and applications should be sent directly to them. Inquire as to whether your school makes such awards.

Lincoln Arc Welding Foundation Scholarships
PO Box 17035
Cleveland, OH 44117
Tel: 216-481-4300

Undergraduate and graduate engineering students interested in arc welding are eligible for these $2,000 scholarships.

Merit Shop Foundation
College Relations Department
729 15th Street, NW
Washington, DC 20005
Tel: 202-637-8800

Students preparing for a career in the construction field who have completed one year of college are eligible for scholarships between $500 and $2,000. Deadline to apply is December 15; write for details.

National Federation of the Blind Scholarships
Scholarship Chair
814 Fourth Avenue, Suite 200
Grinnell, IA 50112
Tel: 515-236-3366

Scholarships of $2,500 are available to blind students in the engineering field. Write for further details.

National Roofing Foundation Scholarships
10255 West Higgins Road, Suite 600
Rosemont, IL 60018-5607
Tel: 708-299-9070

Full-time undergraduate or graduate students enrolled in architecture, engineering, or other program related to the roofing industry are eligible for scholarships of up to $2,000. Scholarships are awarded based on academic standing, recommendations, and interest in a construction industry career. Applications are due January 15; write for information.

National Society of Professional Engineers Scholarships
1420 King Street
Alexandria, VA 22314
Tel: 703-684-2800

Women in the top quarter of their high school class are eligible for $1,000 scholarships to help finance college study in engineering or chemistry.

NSPE State Awards
National Society of Professional Engineers
1420 King Street
Alexandria, VA 22314-2794
Tel: 703-684-2800

To be eligible for this award, applicants must be high school seniors planning to study engineering at a college accredited by the EAC-ABET. They must be U.S. citizens ranking in the upper quartile of their class, with at least a 3.0 GPA and certain minimum scores on the ACT or SAT. The stipend is $1,000 per year; funds are paid directly to the institution rather than to the recipient.

NSPE Student Chapter Member Awards
National Society of Professional Engineers
1420 King Street
Alexandria, VA 22314-2794
Tel: 703-684-2800

To be eligible for this award, applicants must have completed at least two semesters or three quarters of undergraduate studies at a college accredited by EAC-ABET with a minimum GPA of 3.0. U.S. citizenship and student membership in NSPE are required. The stipend is $2,000 per year; funds are paid directly to the institution rather than to the recipient.

Plastics Institute of America Scholarships
277 Fairfield Road, Suite 100
Fairfield, NJ 07004
Tel: 201-808-5950

Scholarships averaging $1,000 are awarded to students enrolled in a plastics technology program; deadline to apply is July 1.

Richard Goolsby Scholarship
Foundation for the Carolinas
PO Box 34769
Charlotte, NC 28234
Tel: 704-376-9541

Engineering students who plan to work in the plastics industry are eligible for this scholarship to help finance their studies. Write for details.

Rust International Corporation Scholarship
National Society of Professional Engineers
1420 King Street
Alexandria, VA 22314-2794
Tel: 703-684-2800

To be eligible for this award, applicants must be high school seniors planning to study engineering at a college accredited by the EAC-ABET. They must be U.S. citizens ranking in the upper quartile of their class, with at least a 3.0 GPA and certain minimum scores on the ACT or SAT. The stipend is $1,000 per year; funds are paid directly to the institution rather than to the recipient.

Samuel Fletcher Tapman Student Chapter/Club Scholarship
American Society of Civil Engineers
1801 Alexander Bell Drive
Reston, VA 20191-4400
Tel: 703-295-6000

Applicants must be ASCE members who are college freshmen, sophomores, or juniors enrolled in a program of civil engineering. Financial need must be demonstrated. Each student chapter of the society may submit an application for only one of its members. The stipend is $1,500.

Science Service Scholarships
1719 N Street, NW
Washington, DC 20036
Tel: 202-785-2255

High school seniors may apply for scholarships ranging from $1,000 to $10,000 per year to study engineering, mathematics, or science. Applications are available from your high school guidance office.

Sheet Metal Workers International Scholarships
1750 New York Avenue, NW, 6th Floor
Washington, DC 20006
Tel: 202-783-5880

Two scholarships are available to members or relatives of members to help fund study in engineering. Applications are due April 1.

Sigma Xi Scientific Research Society Grants
99 Alexander Street
Research Triangle, NC 27709
Tel: 919-549-4691

The society provides grants of up to $1,000 to support student research efforts in the sciences and engineering. Both college undergraduates and graduate students are eligible.

Society for Mining, Metallurgy, and Exploration Scholarships
PO Box 625002
Littleton, CO 80162
Tel: 303-973-9550

SME awards more than 100 scholarships of up to $2,000 each to students in mining engineering, metallurgy, civil engineering, mechanical engineering, geology, or a related field. Write for a copy of the society's guide to scholarships.

Society for the Advancement of Material and Process Engineering Scholarships
PO Box 2459
Covina, CA 91722
Tel: 818-331-0616

College freshmen, sophomores, and juniors are eligible for $1,000 scholarships to help finance study in chemical engineering, material science, chemistry, or physics. The deadline to apply is January 31.

Society of Manufacturing Engineers Scholarships
PO Box 930
Dearborn, MI 48121
Tel: 313-271-1500, ext. 512

The society awards $500 to $2,000 scholarships to manufacturing engineering majors interested in robotics. Applications are due March 1.

Society of Naval Architects and Marine Engineers Scholarships
601 Pavonia Avenue, Suite 400
Jersey City, NJ 07306
Tel: 201-798-4800

Students interested in the study of naval architecture, marine engineering, or ocean engineering may apply for scholarships ranging from $1,000 to $12,000. Applications are due by February 1.

Society of Plastics Engineers Scholarships
14 Fairfield Drive
Brookeville, CT 06804
Tel: 203-775-0471

Scholarships of up to $4,000 are available to students who have finished one year in an engineering college or technical institute and are interested in careers in the plastics industry. Priority is given to student members of the society or to children of members.

SSPI Scholarship
Society of Satellite Professionals International
Educational Award Programs
2200 Wilson Boulevard, Suite 102-258
Arlington, VA 22209

SSPI encourages the entry into the field of satellites and invites students from a wide range of disciplines. The SSPI Scholarship of $1,000 is awarded for special achievement, outstanding research and future promise. Write for eligibility details.

Society of Women Engineers Scholarships
120 Wall Street, 11th Floor
New York, NY 10005
Tel: 212-905-9577

Female engineering students are eligible for scholarships ranging from $1,000 to $2,500. Send a stamped, self-addressed return envelope for complete details.

Tandy Technical Scholars
Texas Christian University
PO Box 32897
Fort Worth, TX 76129
Tel: 817-924-4087

High school seniors are eligible for scholarships of up to $2,500 to help finance study in engineering or science at any college or university. Applicants must be nominated by a high school counselor.

TAPPI Scholarships
TAPPI/Technology Park/Atlanta
PO Box 105113
Atlanta, GA 30348-5113
Tel: 770-209-7222

Must be studying or planning to study engineering or environmental topics and demonstrate significant interest in pursuing career in pulp, paper, and related industries.

Tau Beta Pi/SAE Engineering Scholarship
Society of Automotive Engineers
400 Commonwealth Drive
Warrendale, PA 15096-0001
Tel: 412-772-8534

Applicants must be U.S. citizens who intend to earn a degree in engineering; they must be high school seniors with a 3.75 grade point average and rank in the 90th percentile in both mathematics and verbal on the ACT or SAT. This scholarship is worth $1,000.

Look to the Pros

The following professional organizations offer a variety of materials, from career brochures to lists of accredited schools to salary surveys. Many of them also publish journals and newsletters that you should become familiar with. A number also have annual conferences that you might be able to attend. (While you may not be able to attend a conference as a participant, it may be possible to cover one for your school or even your local paper, especially if your school has a related club.)

When contacting professional organizations, keep in mind that they all exist primarily to serve their members, be it through continuing education, professional licensure, political lobbying, or just "keeping up with the profession." While many are strongly interested in promoting their profession and passing information about it to the general public, these professional organizations are also very busy with other activities. Whether you call or write, be courteous, brief, and to the point. Know what you need and ask for it. If the organization has a Web site, check it out first: what you're looking for may be available there for downloading, or you may find a list of prices or instructions, such as sending a self-addressed, stamped envelope with your requst. Finally, be aware that organizations, like people, move. To save time when writing, first confirm the address, preferably with a quick phone call to the organization itself: "Hello, I'm calling to confirm your address. . . ."

THE SOURCES

■Accreditation Board for Engineering & Technology, Inc.
111 Market Place, Suite 1050
Baltimore, MD 21202-4012
Tel: 410-347-7700

Contact the ABET for a list of accredited engineering schools and programs.

■American Association of Engineering Societies
1111 19th Street, NW, Suite 403
Washington, DC 20036
Tel: 202-296-2237
Web: http://sol.asee.org/aaes

Contact the AAES for information on salaries and their quarterly bulletin, *Engineers*, which discusses careers in engineering.

■American Chemical Society
1155 16th Street, NW
Washington, DC 20036
Tel: 202-872-4600
Web: http://www.acs.org/

The ACS Education Division offers a wide range of resources and services for high school chemistry teachers and students designed to attract students into careers in chemistry. It publishes an award-winning quarterly magazine for high school chemistry students, CHEMMATTERS, which is also available on CD-ROM. The Web site contains a wealth of material, including on-line Career Briefs, which are short narratives illustrating various career options available to students majoring in chemistry, a searchable database of companies offering internships and other types of work experiences in chemistry (including chemical engineering), and information on the U.S. National Chemistry Olympiad. If you are interested in chemistry or chemical engineering, the ACS is a must contact!

■American Indian Science and Engineering Society
AISES Pre-College Programs
5661 Airport Boulevard
Boulder, CO 80301-2339
Tel: 303-939-0023
Email: aisespc@spot.colorado.edu
Web: http://www.colorado.edu/AISES

Through a variety of educational programs, AISES offers financial, academic, and cultural support to American Indians and Alaska Natives interested in science and engineering from middle school through graduate school. It sponsors an essay contest, a national science fair, student chapters of AISES, and summer camps. AISES also offers the *Annual College Guide for American Indians* ($7.50), 96 pages, which describes the top colleges as well as preparation and application information, emphasizing the schools with significant Indian communities and support programs. It lists summer college-prep programs and provides detailed financial information.

American Institute of Aeronautics and Astronautics
370 L'Enfant Promenade, SW
Washington, DC 20024
Tel: 202-646-7400
Web: http://www.aiaa.org

Contact the AIAA for information on careers and education requirements in aeronautical and astronautical engineering; accredited schools; scholarships, awards, and competitions; the *AIAA Student Journal*, and student chapters of AIAA.

American Institute of Chemical Engineers
345 East 47th Street
New York, NY 10017-2395
Tel: 212-705-7660 or 212-705-7338
Web: http://www.aiche.org

Contact the AICE for information on careers in chemical engineering.

American Society for Engineering Education
1818 N Street, NW, Suite 600
Washington, DC 20036
Tel: 202-331-3500
Email: pubsinfo@asee.org
Web: http://www.asee.org/precollegehttp://www.asme.org

ASEE's precollege Web site is a guide for high school students and others interested in engineering and engineering technology careers where you can learn about the different engineering and engineering technology fields, interesting people who got their start as engineers, what engineers actually do, and how to get (and pay for) an engineering education. ASEE offers a variety of inexpensive publications. *An Academic Career: It Could Be for You,* ($2) 12 pages, describes personal and professional rewards of an engineering faculty career. *Thinking of an Academic Career,* ($2) 4 pages, includes employment trends, salary levels, and job hunting suggestions. *Engineering: Your Future,* 20 pages, is available on-line.

Biomedical Engineering Society
PO Box 2399
Culver City, CA 90231
Tel: 310-618-9322
Web: http://bmes.ece.utexas.edu/

Contact BES for a copy of *Planning a Career in Biomedical Engineering,* 8 pages, which outlines typical duties, the specialty areas, employment opportunities, and career preparation.

Institute of Electrical and Electronics Engineers
Pre-College Education Committee
1828 L Street, NW, Suite 1202
Washington, DC 20036
Tel: 202-785-0017
Web: http://ieee.cas.uc.edu/~pcollege/

Contact the IEEE for information on careers in electrical and electronic engineering. Their Web site has a precollege page full of information for high school students.

Institute of Industrial Engineers
Customer Service Center
25 Technology Park/Atlanta
Norcross, GA 30092
Tel: 770-449-0460
Web: http://www.iienet.org

Contact the IIE for information on scholarships and careers in industrial engineering. IIE also offers a variety of free and inexpensive publications. *Industrial Engineering Updates* are 1 to 6 pages each. Titles include *ABET Accredited IE & IET Programs, IE Jobs Today and in the Future, IE Ranks 18th Out of 250 Jobs, Understanding the IE Profession,* and *Salary Survey Results. Planning Your Career as an IE: The People-Orientated Engineering Profession,* 10 pages, covers job demand, work environments, and education and training.

■ **Institute of Transportation Engineers**
525 School Street SW, Suite 410
Washington, DC 20024-2729
Tel: 202-554-8050
Web: http://www.ite.org

Contact the ITE for information on careers in traffic and transportation engineering.

■ **Junior Engineering Technical Society**
Guidance
1420 King Street, Suite 405
Alexandria, VA 22314-2794
Tel: 703-548-5387
Email: jets@nas.edu
Web: http://www.asee.org/jets

Contact JET for information on starting a local student chapter in your high school. JET's Guidance Brochures and Brochures for Most Engineering Specialties (both suitable for middle and high school students) may be reviewed at the Web site. Titles include *Engineering and You, Engineering Is For You, Engineering Technologists and Technicians, Biological Engineering, Electrical Engineering, Environmental Engineering, Mechanical Engineering,* and *Safety Engineering.*

■ **The Minerals, Metals & Materials Society**
Education Department
420 Commonwealth Drive
Warrendale, PA 15086-7514
Tel: 412-776-9000
Web: http://www.tms.org

The MMMS offers a variety of free and inexpensive publications. *Materials Science and Engineering: An Exciting Career Field for the Future,* 4 pages, discusses the future of the field, job opportunities, typical duties, personal qualifications, and education and training. It also lists the 80 accredited colleges and universities offering materials/metallurgical engineering programs. Visit their Web site for information on student chapters, scholarships, and career resources.

■ **National Action Council for Minorities in Engineering, Inc.**
3 West 35th Street, Third Floor
New York, NY 10001-2281
Tel: 212-279-2626
Web: http://www.nacme.org

Contact NACME for information on their Corporate Scholars program and their Engineering Vanguard program. NACME also offers a variety of free and inexpensive publications, including *Academic Gamesmanship: Becoming a Master Engineering Student*, ($1), *Design for Excellence: How to Study Smartly*, ($1), *Financial Aid Unscrambled: A Guide for Minority Engineering Students*, ($1) 24 pages, and *The Sky's Not the Limit*, a pre-college guide to engineering with advice on choosing schools and courses, all available in either English or Spanish.

National Council of Examiners for Engineering and Surveying
PO Box 1686
Clemson, SC 29633-1686
Tel: 864-654-6824
Web: http://www.ncees.org

Contact NCEES for information on licensing, and on their books and pamphlets containing sample PE examination problems and questions. Visit the Web site to read *Engineering As a Career*, written especially for high school students, and to learn more about licensing.

National Society of Black Engineers
1454 Duke Street
PO Box 25588
Alexandria, VA 22313
Tel: 703-549-3207
Web: http://www.nsbe.org/

Contact the NSBE for information on careers in engineering, scholarships, and starting a NSBE Jr. chapter in your high school.

National Society for Professional Engineers
1420 King Street
Alexandria, VA 22314-2794
Tel: 703-684-2800
Web: http://www.nspe.org

Contact the NSPE for information on careers in engineering and on student memberships. Visit the student page at their Web site for a variety of information of interest to high school students considering engineering.

■ **Society of Hispanic Professional Engineers**
5400 East Olympic Boulevard, Suite 210
Los Angeles, CA 90022
Tel: 213-725-3970

Contact the SHPE for information on their competitions and educational pro-grams for engineering students and for information on careers in engineering.

■ **Society of Plastics Engineers**
14 Fairfield Drive
Brookfield, CT 06804
Tel: 203-775-0471
Web: http://www.4spe.org/

Contact SPE for information on careers in plastics engineering.

■ **Society of Women Engineers**
120 Wall Street, Eleventh Floor
New York, NY 10005-3902
Tel: 212-509-9577
Web: http://www.swe.org

SWE offers a variety of free and inexpensive career materials, including *Is Engineering for You?*, 4 pages, which explains college entrance requirements and what women engineers do, and *Guide for High School Women on Becoming an Engineer.* Both of these are available on-line.

■ **TAPPI**
Technology Park
PO Box 105113
Atlanta, GA 30348-5113
Web: http://www.tappi.org

TAPPI offers *An Invitation to Students Interested in Math/Science/Engineering Careers from the Pulp and Paper Industry,* 8 pages, which describes technical career opportunities in the pulp and paper industry and the academic pro-grams that lead to them.

Index